A Finite Element Primer

COPYRIGHT 1992

ISBN 0 903640 17 1

PUBLISHED BY
NAFEMS
BIRNIEHILL
EAST KILBRIDE
GLASGOW, G75 0QU

First published 1986
Reprinted 1987 (with amendments)
2nd Reprint 1991
3rd Reprint 1992

Printed in Great Britain by Bell and Bain Ltd., Glasgow

Contents

Preface

1. Introduction

1.1	Finite Element Method	1
1.2	Matrix Notation	4

2. Structural Analysis

2.1	Fundamentals. Equilibrium. Compatibility. Stress-Strain	9
2.2	The Principle of Virtual Displacements, and Energy Methods	13
2.3	Framework analysis and assembly of **K**	15
2.4	Framework and element stiffnesses	18

3. General Continuum

3.1	Fundamentals again	22
3.2	The P.V.D. again	26

4. Beams

4.1	Exact solution	29
4.2	Finite element solution	32
4.3	Rigid-jointed frameworks	39
4.4	Stiffness transformation	40

5. Two-dimensional Membranes

5.1	Fundamental assumptions	43
5.2	Rectangular elements	47
5.3	Triangular elements	52
5.4	The isoparametric curved quad	56
5.5	Numerical integration	61
5.6	Initial strains	65

CONTENTS

6. Bricks, Plates and Shells — 68

    6.1   Introduction — 68
    6.2   Solid elements — 75
    6.3   Plate bending elements — 79
    6.4   Shell elements

7. Mesh Specification

    7.1   Introduction — 85
    7.2   Geometry specification — 86
    7.3   Mesh generation — 90
    7.4   Mesh density — 94
    7.5   Choice of element type — 102
    7.6   Testing element and mesh suitability — 105
    7.7   Material properties — 112
    7.8   The use of symmetry — 115

8. Assembly and Solution

    8.1   Introduction — 124
    8.2   Checking procedures — 126
    8.3   Element generation and assembly problems — 127
    8.4   Excessive element distortion — 128
    8.5   Incorrect element connections — 131
    8.6   Incorrect mixing of elements — 132
    8.7   Graphical checks — 133
    8.8   Ill-conditioning — 140
    8.9   Solution diagnostics — 141
    8.10 Other program checks and diagnostics — 145
    8.11 Results presentation — 146
    8.12 Improving solution efficiency and accuracy — 147
    8.13 Use of program restarts — 151
    8.14 Substructuring — 152

9. Results Processing

    9.1   Introduction — 154
    9.2   Displacement results — 155
    9.3   Stresses — 156
    9.4   Element stresses — 158
    9.5   Stress averaging — 159
    9.6   Methods of stress averaging — 160
    9.7   Stress presentation — 162

9.8	Element strains	164
9.9	Calculation of reaction forces	165
9.10	Graphical presentation of results	166
9.11	Using the results to refine the mesh	168

10. Dynamics

10.1	Introduction	170
10.2	Undamped free vibration	173
10.3	Modelling considerations for dynamic analysis	180
10.4	Forced reponse	184
10.5	Methods for calculating forced response	186
10.6	Damping idealisation	190
10.7	Condensation and dynamic substructuring	196
10.8	Primary and secondary components	200
10.9	Number of nodes for dynamic analysis	202
10.10	Calculation of dynamic stresses	206
10.11	Results recovery	207
10.12	Structural modifications	208
10.13	Wave propagation	209
10.14	Seismic analysis	210
10.15	Random vibrations	212

11. Nonlinear Analysis

11.1	Introduction	215
11.2	Gross deformation	217
11.3	Incremental solutions	224
11.4	Inelastic material behaviour	227

12. Modelling

12.1	Introduction	232
12.2	Basic	232
12.3	Linear and non-linear problems	233
12.4	Joints	234
12.5	Offsets	235
12.6	Supports	239
12.7	The use of constraint equations	242
12.8	Comparison of the forms of constraints	247
12.9	Relationship between the forms of constraints	248

CONTENTS

12.10	Using mixtures of element types	249
12.11	Modelling material properties	251
12.12	Loadings	257
12.13	Modelling considerations including loading effects	261

13. Other Field Problems

13.1	Introduction	262
13.2	The method of weighted residuals	263
13.3	The heat conduction problem	265
13.4	Other field problems	270

References 272

Index 273

Preface

This Primer is one of a series of publications produced by NAFEMS, an organisation funded by the Department of Trade and Industry, and charged with maintaining standards and quality assurance in Finite Element technology in the U.K. This brief is a very broad brush, and the agency intends to tackle the problem of standards in several ways. The most obvious goal is to encourage quality assurance in commercially available finite element systems themselves, and this will be done by prescribing minimum acceptable standards, desirable features, and a series of tests or procedures against which systems should be measured and their performance disseminated in NAFEMS publications.

But there is another problem. Many of today's finite element systems are intended to be robust and friendly; and may be used by engineers and scientists who are unfamiliar with the finite element method, its virtues and its vices. Indeed some finite element systems may be embedded in a CAD/CAE package and the user may be a production engineer, designer, or systems analyst who not only does not understand the characteristics of finite element systems but may very well not even wish to know. The Agency believes this situation to be dangerous and has therefore taken up the task of encouraging and improving finite element education in several ways. A detailed register will be kept of all known educational sources, whether they be short courses or full-time, state or private. The **Guidelines to Finite Element Practice** has already been published and will be kept updated. Finally the problems of the novice have been identified and addressed in this Primer. The aim is not to write another finite-element text book. There are scores of these and many are daunting for the new reader since, not unnaturally, many go into some algebraic detail of specific elements or they highlight areas of current research. The aim is also not to write a detailed instruction manual, since this cannot be done without referring to a specific system. Indeed all commercial systems have their own introductory and users' manuals, together with more advanced programmers' manuals in some cases, and these are usually very good.

PREFACE

They do not however discuss the vices or failings of the system, for obvious reasons.

This Primer will try to explain the basis of the Finite Element Method, stressing the essentially simple fundamental concepts without digressing into lengthy algebra or oversmart mathematics. There will be no functional analysis in Hilbert spaces, and Lagrange multipliers will just be mentioned even though they are used sparingly. Some algebra is inevitable, otherwise the text becomes a descriptive routine rather like explaining the fitting together of pieces of a jig-saw, and this can lead to all sorts of confusions. However, the algebra can be kept concise using matrix notation which is obligatory in discussing these methods. The Primer will strip some of the mysteries from the method and in particular will explain precisely what the method does exactly and what it does approximately, and as a consequence which errors are important and which are not. Some errors are very useful sources of guidance.

Following fundamental chapters into the nature of the method for static problems and various types of structures, the nature of proprietary finite element systems will be discussed, together with those features which the user has the right to expect. The latter parts of the Primer will extend the method to dynamic problems, non-linear elasto-plastic and buckling problems, heat transfer, and in Chapter 12 examples will be given to illustrate some of the modelling difficulties encountered in making a finite element idealisation in the first place.

It will be assumed that the reader has a basic understanding of applied mechanics and that differentiation or integration are not completely foreign concepts. The validity of a material's constitutive laws will not be questioned. The opening section even runs through a quick introduction to matrices and their manipulation – we realise that nowadays most engineers are familiar with matrix algebra but it does also give us an opportunity gently to introduce the notation used in this text.

In most chapters the fundamentals behind the finite element method are very briefly discussed. It is possible of course to omit this, and simply state the equations which are necessary to formulate the numerical models. However it has been shown that a grasp of the fundamentals is useful in judging errors, choosing idealisations, and even understanding the user-manuals! Such introductions can be omitted if familiar. If any reader therefore finds some parts of the Primer too elementary we apologise. If any find parts a little demanding, then perhaps the balance is about right.

Notation

General

Displacements $\mathbf{u} = [u, v, w]$
Co-ordinates (x, y, z)
Nondimensional (mapped) co-ordinates $(\zeta_1, \zeta_2, \zeta_3)$
Body forces $\mathbf{p}_v^t = [p_{vx} \quad p_{vy} \quad p_{vz}]$
Surface forces $\mathbf{p}_s^t = [p_{sx} \quad p_{sy} \quad p_{sz}]$
Strain $\boldsymbol{\varepsilon}^t = [\varepsilon_{xx} \quad \varepsilon_{yy} \quad \varepsilon_{zz} \quad \varepsilon_{xy} \quad \varepsilon_{yz} \quad \varepsilon_{zx}]$
Stress $\boldsymbol{\sigma}^t = [\sigma_{xx} \quad \sigma_{yy} \quad \sigma_{zz} \quad \sigma_{xy} \quad \sigma_{yz} \quad \sigma_{zx}]$
Material stiffness $\mathbf{E}$: $\boldsymbol{\sigma} = \mathbf{E}\boldsymbol{\varepsilon}$
Material stress-strain $\boldsymbol{\varepsilon} = \mathbf{f}\boldsymbol{\sigma} + \boldsymbol{\eta}$

Element

Displacements $\mathbf{d}_g$, forces $\mathbf{P}_g$
Interpolations (shape functions) $\mathbf{u} = \mathbf{N}\mathbf{d}_g$
Compatibility $\boldsymbol{\varepsilon} = \partial \mathbf{u}, = \partial \mathbf{N}\mathbf{d} = \mathbf{B}\mathbf{d}$
Stiffness $\mathbf{k} = \int_v \mathbf{B}^t \mathbf{E} \mathbf{B} \, dv$; forces $\mathbf{P} = \mathbf{k}\mathbf{d}$

Global

Displacements $\mathbf{r}$, forces $\mathbf{R}$
Stiffness $\mathbf{K}$. $\mathbf{R} = \mathbf{K}\mathbf{r}$
Flexibility $\mathbf{F}$. $\mathbf{r} = \mathbf{F}\mathbf{R}$

1. Introduction

1.1 The Finite Element Method

It would be a brave and foolhardy person who credited a single individual with the invention of anything, particularly so the finite element method, which has now passed into common usage in all branches of engineering and science and must be judged a huge success story in both intellectual and commercial terms. The idea was undoubtedly used by the renowned mathematician Courant in 1943, but he published in the Bulletin of the American Mathematical Society and it was probably considered an idle diversion by classical analysts and totally ignored by practising engineers who of course are not generally diligent readers of mathematics journals. It was not until the early fifties when, almost simultaneously in the U.K. and in California, researchers began to exploit a method which relied totally on the infant digital computer to set up the problem as well as solve it. Almost all of the pioneers had connections with the aircraft industry since this sector had the strongest motives for accurately analysing minimum structures of great complexity. It also had the necessary research capital needed to buy these early digital computers which were notoriously expensive and unreliable collections of thermionic valves and relays. They (the computers) were also very unfriendly to address. It was because of the problem of addressing machines and describing a geometrically complex structure that *matrix notation* became the lingua franca. Indeed during the 1950s the methods were known as 'Matrix Methods' and most of the protagonists described the techniques rather like an extension of framework analysis, and not like a piecewise Rayleigh–Ritz process. The use of matrix descriptions is not of course confined to structures or to linear algebra, or to computational methods, but the language did make a powerful impact on the way that structural engineers approached large-scale analytical problems. A resume of this language appears in the next section.

The reasons for needing the finite element method and its popularity can be appreciated if we imagine the problem of solving the stresses and strains in structures in almost any branch of engineering such as mechanical, civil, aeronautical, marine, nuclear, and so on. Most modern structures are complicated in the extreme. Sometimes of course a small component (or element) of a structure may appear to have a simple form like a beam in a framework or a thin plate in an automobile. The analysis of such simple components forms the subject of elementary textbooks in strength of materials, and as often as not they lead to the solution of a differential equation, in one or two variables, which can be solved in some form. But most modern structures are not that simple – they may be extremely complicated creations in three dimensions. Even if they are an assembly of simple beams or 'pinended' bars then the local forces in those beams still have to be isolated *before* using textbook formulae. Thus the main structural problem is essentially one of geometrical shape in the first instance. (This is in contrast to fluids where many shapes are simple but the flow rarely so.) Once the computer has been used to describe and manipulate the complex geometry, then the laws of equilibrium, motion, strain compatibility and stress-strain can be invoked and the process of analysis begun.

Computers can be instructed to describe some complete structures in terms of continuous functions, or to solve differential equations in terms of continuous analytical functions, but this is not a feasible option for a Boeing 747 or for a high rise block of apartments. We turn instead to a *discretised approach* where the geometric shape or the internal stress-strain-displacement fields are described by a series of discrete quantities (like coordinates) scattered over the surface or distributed through the interior of the structure. Computers are ideally equipped for storing long lists of separate numbers and manipulating them.

The basic idea of the finite element method is really quite simple, and may be visualised as doing two entirely separate things. Firstly a structure is divided (hypothetically) into finite elements which are so small that the *shape* of the displacement or stress field can then be approximated without too much danger, leaving only the *magnitude* to be found. (In this text we concentrate on assuming displacement shapes rather than stress since this is by far the most popular approach.) The problem is thereby discretised, that is we are left only with having to find a set of numbers. The 'shapes' may be polynomials, trigonometric functions and so on, as we shall see.

Secondly, all the individual elements have to be assembled together in such a way that the displacements and stresses are continuous in some

INTRODUCTION

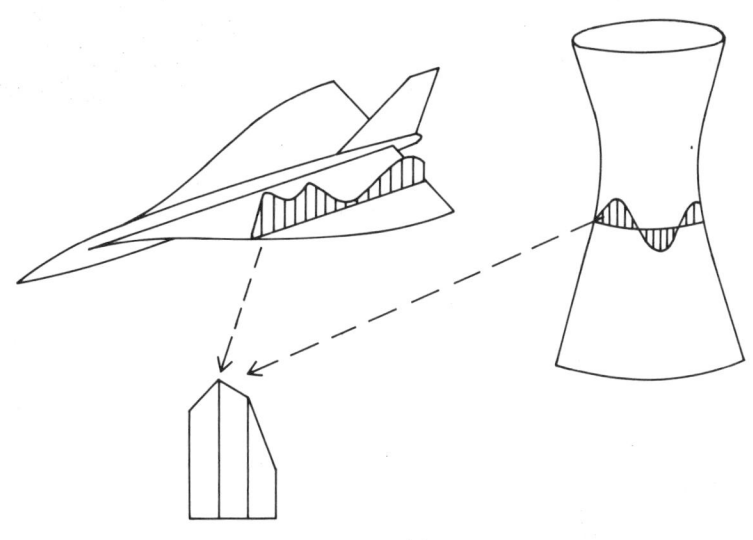

Figure 1.1

fashion across element interfaces, the internal stresses are in equilibrium with each other and the applied loads, and the prescribed boundary conditions are satisfied.

The first part of the finite element process is fundamental and involves choosing the correct and appropriate types of elements and describing and evaluating their properties. These aspects are discussed later in some detail. Although modern finite element systems do protect users from having to evaluate element properties, it does help to be able to understand and exercise judgement when modelling the structure, and spotting wrong answers due to using inadequate elements. It also helps to understand the pedigree of elements when deciding whose finite element system to try!

The second part of the process is the assembly of the elements and then solving the complete structure and so on. The proprietary system will do this of course, but again it pays to know what is involved since sometimes this process breaks down, or it simply becomes an inefficient process because the structure has been described inconveniently.

Fortunately we can separate the above two features by considering simple frameworks first where the elements are real distinct components requiring

A FINITE ELEMENT PRIMER

no approximation at all. Assembly and solution are everything. But first a description of the language.

1.2 Matrix Notation

We have already implied that we wish to describe the behaviour of a structure by a large number of discrete displacements at chosen points. These points may be natural choices like the joints in a framework, but often they will be 'nodes' in a hypothetical grid over the surface or through the volume. A single displacement, r, at a point could be labelled r_i where it is understood that the subscript 'i' may be one of many, say $i = 1, 2, \ldots, i, \ldots, n$. We assemble this best as a *column matrix* having n rows thus:

$$\begin{bmatrix} r_1 \\ r_2 \\ r_3 \\ \vdots \\ r_i \\ \vdots \\ r_n \end{bmatrix} \quad (1.1)$$

Sometimes this is referred to as a *column vector* having n components. It will be necessary to adopt a form of convenient shorthand instead of writing columns like (1.1) every time. One way would be to use a symbol (r) followed by an 'argument' [i] in brackets so that (1.1) becomes

$$r[i], \; i = 1, 2, 3, \ldots, n. \quad (1.2)$$

This is the form usually adopted when telling a computer what to do with a list of numbers like 'read $r[i]$' or 'print' or 'multiply' etc. by simply writing a single statement and looping $i = 1, \ldots, n$. However this text would look a little daunting if we kept to Fortran-type symbols, so we will adopt the usual convention and use bold type for matrices, so a list of displacements will be written as

$$\mathbf{r} = r[i] = \begin{bmatrix} r_1 \\ r_2 \\ \vdots \\ r_n \end{bmatrix}. \quad (1.3)$$

INTRODUCTION

Suppose, at the chosen nodes of the structure, there are also a series of concentrated forces $R_1, R_2, R_3, \ldots, R_i, \ldots, R_n$, all acting in the same sense as the labelled displacements **r**. These force and displacement components are then said to *correspond*. Now if the problem is *linear*, that is all displacements, stresses and strains are proportional to the applied loads, then as we increase the forces from zero to their final values the total work done by all of these forces will be

$$W = \tfrac{1}{2}R_1 r_1 + \tfrac{1}{2}R_2 r_2 + \cdots + \tfrac{1}{2}R_n r_n. \tag{1.4}$$

This work, in the absence of dynamic effects, is absorbed by the structure as elastic potential energy, or 'strain energy'. We should like to summarise (1.4) in a convenient way, and one possibility is to arrange the forces as a *row matrix*. Then we adopt the standard convention that a row matrix can multiply a column matrix taking one element product at a time, and then summed. (The row × column rule.)

$$[R_1, R_2, R_3, \ldots, R_i, \ldots, R_n] \begin{bmatrix} r_1 \\ r_2 \\ \vdots \\ r_i \\ \vdots \\ r_n \end{bmatrix} = R_1 r_1 + R_2 r_2 + \cdots + R_n r_n = \sum_i R_i r_i.$$

However it is not at all convenient to have to describe forces and displacements differently by rows and columns so we adopt the same column form for forces:

$$\mathbf{R} = \begin{bmatrix} R_1 \\ R_2 \\ R_3 \\ \vdots \\ R_n \end{bmatrix}. \tag{1.5}$$

To turn this into a row the matrix is said to be 'transposed' (columns become rows) and is identified by a superscript '*t*'.

$$\mathbf{R}^t = [R_1, R_2, R_3, \ldots, R_i, \ldots, R_n].$$

Equation (1.4) now simply becomes

$$W = \tfrac{1}{2}\mathbf{R}^t\mathbf{r}. \tag{1.6}$$

We now go a stage further and assume that in our linear structure all displacements **r** have been solved in terms of **R** to which they are proportional. Denoting these coefficients of proportionality as F, we could write

$$r_1 = F_{11}R_1 + F_{12}R_2 + F_{13}R_3 + \cdots + F_{1n}R_n.$$

The symbol F is obviously a measure of *flexibility* and two subscripts are necessary since there are other displacements as well:

$$r_2 = F_{21}R_1 + F_{22}R_2 + F_{23}R_3 + \cdots + F_{2n}R_n$$

and finally

$$r_n = F_{n1}R_1 + F_{n2}R_2 + \cdots + F_{nn}R_n.$$

Having summarised displacements and forces as column matrices it is most convenient to summarise the above equations as

$$\mathbf{r} = \mathbf{FR} \tag{1.7}$$

which is shorthand for:

$$\begin{bmatrix} r_1 \\ r_2 \\ \vdots \\ r_n \end{bmatrix} = \begin{bmatrix} F_{11} & F_{12} & \cdots & F_{1n} \\ \vdots & & & \vdots \\ \vdots & & & \vdots \\ F_{n1} & F_{n2} & \cdots & F_{nn} \end{bmatrix} \begin{bmatrix} R_1 \\ \vdots \\ \vdots \\ R_n \end{bmatrix}. \tag{1.8}$$

provided that the 'row × column' rule is followed for each individual row of the *array* **F**. This array

$$\mathbf{F} = F[i, j], \quad \text{for } i = 1, 2, \ldots, n;\ j = 1, \ldots, n$$

is known as the *Flexibility Matrix*.

These very simple matrix operations will take us far and in many cases will enable us to discuss structural behaviour as if the structure had simply one degree of freedom like a single spring.

INTRODUCTION

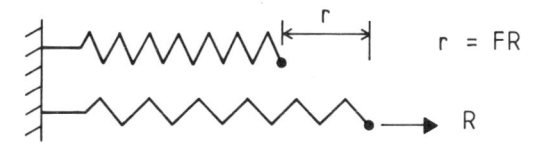

Figure 1.2

This idea of discussing all structures as if they were a simple spring is tempting, particularly when we come to dynamic problems with inertia and damping. But we must be a little careful and not treat matrices in quite the same cavalier fashion as we treat simple numbers (scalars). This is particularly so when executing several multiplications since the number of rows and columns must always match. We might for example have a matrix **C** of p rows and q columns, in the middle of a product:

$$\underset{(1 \times n)}{\mathbf{A}} = \underset{(1 \times p)}{\mathbf{B}} \times \underset{(p \times q)}{\mathbf{C}} \times \underset{(q \times n)}{\mathbf{D}}.$$

Notice how the columns and rows of successive matrices in the product must match. Either of the products **BC** or **CD** can be executed first to form a $(1 \times q)$ or a $(p \times n)$ matrix, before the final multiplication forms the $(1 \times n)$ column matrix. However it is clear that $\mathbf{D}(q \times n)\mathbf{C}(p \times q)\mathbf{B}(1 \times p)$ is not the same and cannot even be executed. Matrix products, as an early scholar said, should be likened to putting on socks and shoes. The correct order may be crucial (and can be verified experimentally).

It can readily be verified that the transpose of a product is the product of the transposed matrices, that is

$$(\mathbf{ABC})^t = \mathbf{C}^t\mathbf{B}^t\mathbf{A}^t.$$

If we apply this to (1.6) we know that work is a simple scalar and so is equal to its transpose.

Thus

$$\tfrac{1}{2}\mathbf{R}^t\mathbf{r} = W = \tfrac{1}{2}(\mathbf{R}^t\mathbf{r})^t = \tfrac{1}{2}\mathbf{r}^t\mathbf{R}$$

$$= \tfrac{1}{2}r_1 R_1 + \tfrac{1}{2}r_2 R_2 + \cdots + \tfrac{1}{2}r_n R_n$$

which is obvious. But if we employ the equation (1.7) for **r** then

$$\mathbf{R}^t\mathbf{r} = \mathbf{R}^t\mathbf{F}\mathbf{R}$$

7

whereas

$$r^t R = R^t F^t R.$$

For these to be the same, for any set of loads applied in any order, F^t must be the same as F, which means that the flexibility matrix must be *symmetrical* about the leading diagonal ($F_{ij} = F_{ji}$).

Finally, anticipating the forms of the equations produced by the finite element displacement method, we shall derive a set of equations relating r to R in the form

$$Kr = R \qquad (1.9)$$

where K is this time clearly a measure of *stiffness*. (Increase K and bigger loads are required for the same deflection.) The matrix K is known as the *stiffness matrix*. It is also symmetrical following the same arguments used for F. We shall have to solve equations like (1.9) for r before we can find internal stresses and strains; and the solution of (1.9) is often written symbolically as

$$r = K^{-1} R \qquad (1.10)$$

where $(K)^{-1}$ means the 'inverse' of K and clearly is the same as F. The inverse is often written in texts on Linear Algebra as

$$K^{-1} = \frac{\text{Adj } K}{|K|} \qquad (1.11)$$

where $|K|$ is the determinant of K. We will not pursue this since in finite element formulations it is usual to solve (1.9) without forming the full inverse – a topic to be discussed later.

For readers totally familiar with matrix manipulation this diversion is now concluded, and we turn to the serious business of Structural Analysis.

2. Structural Analysis

2.1 The Fundamentals

In solving any structural problem, whatever the type of structure, whatever the type of loading, be it static or dynamic, and whatever the nature of the structure's material, there are only three types of argument we can deploy. These three arguments are quite separate and distinct, and it is important to understand their simplicity, especially when delving into the intricacies of the finite element method where the wood is often obscured by the trees. The arguments are:

i. *Equilibrium*
   These arguments relate stress (σ) to applied forces, or often stresses to other stresses whether there are applied forces or not. If the structure is excited dynamically then the 'inertia forces' can be inserted into the equations of equilibrium as if the problem was still static. If displacements are small then the equations of equilibrium are linear.

ii. *Compatibility*
   These relate strains (ε) to displacements and are purely geometrical arguments which depend on the definition of strain and the type of deformation and geometry of the particular structure. If the displacements are small then the compatibility equations are also linear.

iii. *Stress-strain Law*
   These 'constitutive relationships' are empirical and depend on experimental evidence. They may include thermal effects, and for ferrous materials the relationship may be *elasto-plastic* with irreversible plasticity. For many structural materials within their useful working range these laws may be taken as linear.

To illustrate the analytical process at work let us consider the trivial problem of the 3-bar pinjointed framework shown in Figure 2.1.

A FINITE ELEMENT PRIMER

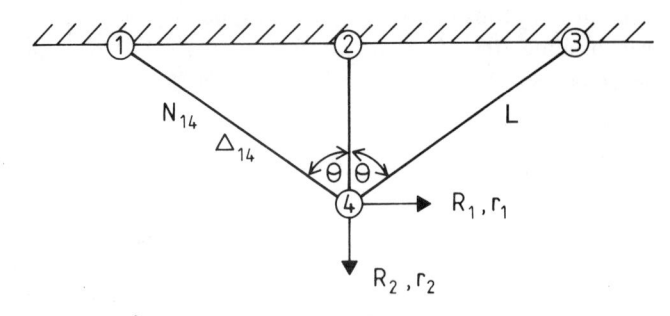

Figure 2.1

The joint (4) is subjected to two forces R_1, R_2, and consequently displaces by r_1 and r_2. Denote the internal bar forces by N and the bar extensions by Δ.

First of all the bar elements:

If the bars are pinended and have uniform cross-sectional area A, then the stress in a bar is given by an equilibrium equation like

$$A\sigma = N. \tag{2.1}$$

The bar compatibility equation is simply

$$\varepsilon = \Delta/l. \tag{2.2}$$

Now the joint (4) which is acted upon by the three bars:

The equilibrium equations are:

$$N_{14}\sin\theta - N_{34}\sin\theta = R_1,$$
$$N_{14}\cos\theta + N_{24} + N_{34}\cos\theta = R_2. \tag{2.3}$$

The compatibility equations relate a bar extension Δ to the displacements of the ends of the bar. Taking these displacements to be so small that only components along the bar will stretch it, we have

$$\Delta_{14} = r_1\sin\theta + r_2\cos\theta,$$
$$\Delta_{24} = r_2, \tag{2.4}$$
$$\Delta_{34} = -r_1\sin\theta + r_2\cos\theta.$$

The stress-strain relationships we assume to be 'Hooke's Law', and in addition we assume that a temperature increase T will produce a proportional increase in strain αT which can be added to that due to stress, viz:

$$\varepsilon = \frac{\sigma}{E} + \alpha T. \qquad (2.5)$$

Note that ε is the total strain and is the same strain as in the compatibility equations. The stress in (2.5) is the total stress and no attempt is made to split the stress into 'that due to load' and 'that due to temperature'. This concept is followed in some texts but only confusion can arise if this idea is carried through to two- and three-dimensional solids.

There is nothing more to be added to equations (2.1) through (2.5), all the information is there. We quickly proceed to a *displacement solution* as follows. Suppose only bar (2)-(4) is heated. Then using (2.1), (2.2) and (2.5) we can express Δ in terms of N thus

$$\frac{\Delta_{14}}{L} = \frac{N_{14}}{AE}; \quad \frac{\Delta_{24}}{L\cos\theta} = \frac{N_{24}}{AE} + \alpha T; \quad \frac{\Delta_{34}}{L} = \frac{N_{34}}{AE}. \qquad (2.6)$$

Equation (2.6) can be used to convert the compatibility equations (2.4) to force $\sim$ displacement equations as follows

$$N_{14} = \frac{AE}{L}(r_1 \sin\theta + r_2 \cos\theta),$$

$$N_{24} = \frac{AE}{L\cos\theta} r_2 - AE\alpha T, \qquad (2.7)$$

$$N_{34} = \frac{AE}{L}(-r_1 \sin\theta + r_2 \cos\theta).$$

All bar forces are now known in terms of just the two joint displacements. But there are just two *corresponding* joint equations of equilibrium, so we know a solution is possible. Substituting then (2.7) into (2.3) we obtain the two equations:

$$\frac{2AE}{L}(\sin^2\theta)r_1 = R_1,$$

$$\frac{AE}{L}(2\cos^2\theta + \sec\theta)r_2 = R_2 + AE\alpha T.$$

Notice that these equations are the same form as (1.9), $\mathbf{Kr}=\mathbf{R}$, except that the right hand side has a thermal loading term added. Having solved for r_1 and r_2 it is straightforward to back-substitute into (2.4) to obtain bar extensions, and hence strains from (2.2) and stresses from (2.5).

This simple solution is an example of the *displacement method*, so-called because the displacements are those variables which are solved first. It illustrates all the attendant advantages of being absolutely straightforward and methodical, and it works whether the structure is statically determinate or not. Once the initial assumptions have been made about the nature of the structure and its strain fields, the displacement method should be predictably straightforward.

Structural analysis can therefore be summarised in the schematic of Figure 2.2.

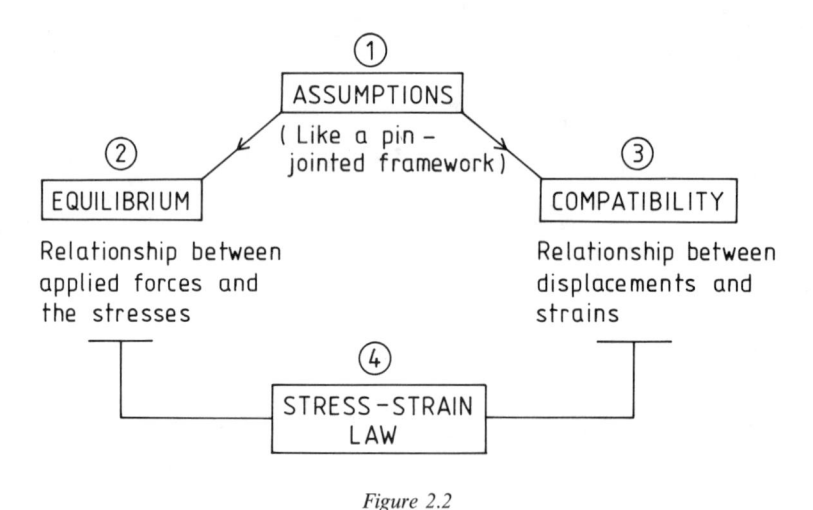

Figure 2.2

The route followed in the displacement method is (1)-(3)-(4)-(2). If the structure is statically determinate (very few are) then (1)-(2) is the natural route for stresses, followed by (4)-(3) only if the displacements are needed. If the structure is statically indeterminate it is only possible to satisfy equilibrium in terms of unknown redundancies which are then found by completing the route (1)-(2)-(4)-(3). Note that the stress-strain law, however complicated, has nothing whatsoever to do with equilibrium (2) or compatibility (3) arguments, with which we shall be most concerned. It is a common misconception that thermal strains for example should modify (3).

In practice it is straightforward to extend our illustrated displacement method to pinjointed frameworks with any number of joints in a systematic fashion. However the easiest way of doing this is to utilise another concept – the Principle of Virtual Work. It will also be found that this technique is just as powerful for rigid jointed frames, beams, plates and shells, and in fact for any solid. Moreover – and this is the vital feature – the principles of virtual work enable us to use *approximations* which are the heart of the finite element method.

2.2 The Principle of Virtual Displacements and Energy Methods

We now illustrate a technique, initially for the pinjointed framework, which is an indirect way of enforcing either equilibrium or compatibility. It is a genuine alternative, and one may ask why use an alternative? We hope the advantages become obvious.

The aim is to replace the equation of equilibrium by a *Work* argument. The stress-strain law is to be irrelevant to this process, indeed it might be nonlinear for example. It is clear then that we cannot use *real* work or strain energy which must depend on this law. We therefore simply use the product Rr as a *virtual work* instead of the true work

$$W = \int_0^r R\,dr$$

as depicted in Figure 2.3.

Real Work · Virtual Work

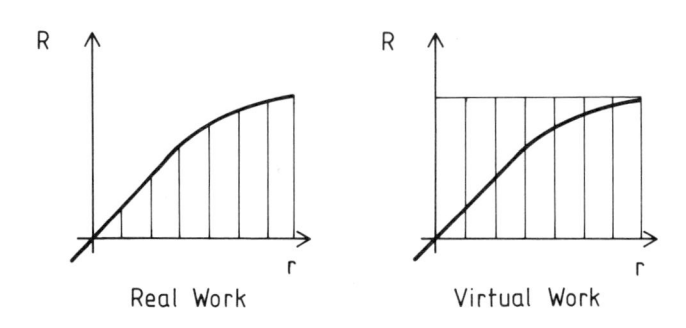

Figure 2.3

In the linear case of course $W = \frac{1}{2}Rr = \frac{1}{2}$(Virtual Work). The above product is the work done by an applied force. In a similar fashion the internal virtual work (or virtual strain energy) stored in the bars of a pinjointed framework is $N\Delta$.

We now go a stage further (and the reasons will become clear) and consider *virtual displacements* $\bar{r}$ and bar extensions $\bar{\Delta}$ which are arbitrary and not related to either the actual displacements or to the forces producing them. In other words we will feel free to *allow* some convenient displacement pattern and examine the virtual work done by forces and stresses over these displacements and strains. These virtual displacements and strains will be denoted by a 'bar', thus $\bar{r}, \bar{d}, \bar{\Delta}, \bar{\varepsilon}$.

The *Principle of Virtual Displacements* (PVD from now on) simply *equates internal work to external work* using the products of real forces and virtual displacements. We prove the sanity of this curious method later, but first examine the case of our pinjointed frame. Here we have (Figure 2.1)

$$N_{14}\bar{\Delta}_{14} + N_{24}\bar{\Delta}_{24} + N_{34}\bar{\Delta}_{34} = R_1\bar{r}_1 + R_2\bar{r}_2. \quad (2.8)$$

Equation (2.8) is not a promising beginning since it gives no information at all. But now let us insist that these *virtual displacements and strains satisfy compatability* (2.4). On inserting these conditions for $\bar{\Delta}$ into (2.8) and collecting coefficients of $\bar{r}_1$ and $\bar{r}_2$ we obtain

$$\bar{r}_1(N_{14}\sin\theta - N_{34}\sin\theta - R_1) + \bar{r}_2(N_{14}\cos\theta + N_{24} + N_{34}\cos\theta - R_2) = 0. \quad (2.9)$$

The subtlety of using virtual displacements now emerges. Since $\bar{r}_1$ and $\bar{r}_2$ are unrelated to each other we could put either to zero. Equation (2.9) must be zero whatever the values of $\bar{r}_1$ and $\bar{r}_2$ and this can only be true if their coefficients vanish. Equation (2.9) therefore delivers

$$\bar{r}_1 \rightarrow \qquad N_{14}\sin\theta - N_{34}\sin\theta - R_1 = 0,$$

$$\bar{r}_2 \rightarrow \qquad N_{14}\cos\theta + N_{24} + N_{34}\cos\theta - R_2 = 0.$$

These are indeed the equations of equilibrium again. The PVD will deliver as many equations of equilibrium as there are separate displacement components.

It is also possible to switch the 'bars' from displacements to forces and obtain a *Principle of Virtual Forces* which will deliver the equations of compatibility provided the equations of equilibrium are enforced. Based on this a finite element method can be developed which was known as the Matrix Force Method in the 1950s. It was popular then because the number of unknowns (redundancies) is small and computers were expensive. It requires more skill to apply than the PVD and its popularity has lapsed, although it is used in some modern finite element systems in the guise of 'equilibrium elements' whose flexibility is inverted to become a stiffness, and so emulate a displacement element (see refs. 1 and 2).

Although we shall use virtual work arguments throughout this Primer, many textbooks and finite element systems manuals use *The Principle of Minimum Potential Energy* as an equivalent statement. The potential energy U consists of two parts: the *internal* (Strain Energy) (U_i), which in a simple bar obeying Hooke's law would be

$$U_i = \tfrac{1}{2} N \Delta = \frac{AE}{2l} \Delta^2.$$

The *external* potential energy (U_e) is defined as usual as

$$U_e = -\sum_i R_i r_i$$

(so for example a gravitational force $R = -mg$ acts in the opposite sense to a vertical displacement $r = h$, and the potential becomes $U = mgh$).

The energy principle simply states that $U = U_i + U_e$ is a minimum for stable equilibrium. So for example in our frame of Figure 2.1

$$U = \frac{AE}{2l}(r_1 \sin\theta + r_2 \cos\theta)^2 + \frac{AE}{2l\sin\theta} r_2^2 + \frac{AE}{2l}(-r_1 \sin\theta + r_2 \cos\theta)^2$$

$$- R_1 r_1 - R_2 r_2.$$

Putting $\partial U/\partial r_1 = 0$ and $\partial U/\partial r_2 = 0$ will deliver the same equilibrium equations as (2.3).

2.3 Framework Analysis and Assembly of Stiffness Matrix

The way is now clear to apply the PVD to a framework consisting of many bars, a typical 'g'th bar being shown in Figure 2.4.

A FINITE ELEMENT PRIMER

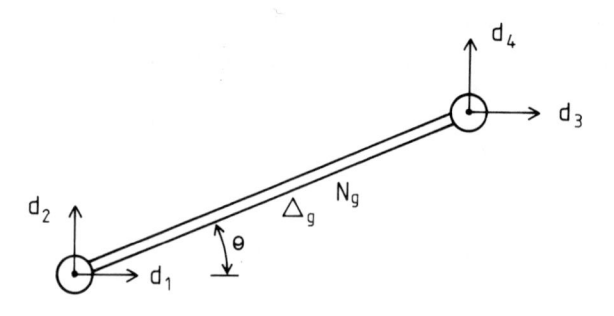

Figure 2.4

Because this bar is a typical element it is convenient to label its displacements as

$$\mathbf{d}_g = \begin{bmatrix} d_1 \\ d_2 \\ d_3 \\ d_4 \end{bmatrix}$$

which will be a known selection of the total (*global*) list **r**. This fact is represented symbolically by writing for the '*g*'th element,

$$\mathbf{d}_g = \mathbf{a}_g \mathbf{r}. \qquad (2.10)$$

The selection matrix $\mathbf{a}_g$ in this case is simply four rows with four unit elements and many zeros.

The PVD for a complete framework of many elements is simply

$$\sum_g N_g \bar{\Delta}_g = \mathbf{R}^t \bar{\mathbf{r}} \qquad (2.11)$$

where the summation is carried out over $g = 1, 2, 3, \ldots$. Again this gives no information until we enforce compatibility which for the general bar is

$$\Delta_g = (d_3 - d_1)\cos\theta + (d_4 - d_2)\sin\theta$$

or

$$\Delta_g = \mathbf{B}\mathbf{d}_g \qquad (2.12)$$

where

$$\mathbf{B} = [-\cos\theta \quad -\sin\theta \quad \cos\theta \quad \sin\theta]. \tag{2.13}$$

Now $N_g = AE\Delta_g/l = k\Delta_g$; where $k = AE/l$ is a measure of the bar's stiffness. The element virtual work can therefore be written as

$$N_g \bar{\Delta}_g = \Delta_g k \bar{\Delta}_g = \mathbf{d}_g^t \mathbf{B}^t k \mathbf{B} \bar{\mathbf{d}}_g \quad \text{using (2.12)}.$$

So (2.11) becomes

$$\sum_g \mathbf{d}_g^t \mathbf{k}_g \bar{\mathbf{d}}_g = \mathbf{R}^t \bar{\mathbf{r}} \tag{2.14}$$

where the symmetric matrix

$$\mathbf{k}_g = \mathbf{B}^t k \mathbf{B} \tag{2.15}$$

is shortly shown to be a genuine element stiffness. Putting (2.10) into (2.14),

$$\mathbf{r}^t \sum_g \mathbf{a}_g^t \mathbf{k}_g \mathbf{a}_g \bar{\mathbf{r}} = \mathbf{R}^t \bar{\mathbf{r}}$$

or

$$[\mathbf{r}^t \mathbf{K} - \mathbf{R}^t] \bar{\mathbf{r}} = 0$$

where

$$\mathbf{K} = \sum_g \mathbf{a}_g^t \mathbf{k}_g \mathbf{a}_g. \tag{2.16}$$

For the above expression to be zero for all $\mathbf{r}$ each row must separately vanish in the square brackets. Therefore $\mathbf{R}^t - \mathbf{r}^t \mathbf{K} = \mathbf{0}$, or

$$\mathbf{Kr} = \mathbf{R}. \tag{2.17}$$

We have therefore derived the global set of displacement equations, and the symmetric matrix $\mathbf{K}$ in (2.17), given by (2.16), is the global stiffness matrix. Having shown algebraically how to assemble a stiffness matrix it is instructive to look in detail at some of the steps.

Firstly compare the two work expressions for the complete structure

$$\mathbf{R}^t \bar{\mathbf{r}} = \mathbf{r}^t \mathbf{K} \bar{\mathbf{r}} \quad \text{and} \quad = \sum_g \mathbf{d}_g^t \mathbf{k}_g \bar{\mathbf{d}}_g.$$

A FINITE ELEMENT PRIMER

Consider the Ith and Jth rows in $\mathbf{r}$, say r_I and r_J. The term K_{IJ} situated at the Ith row and Jth column of $\mathbf{K}$ will clearly emerge in the work product as

$$r_I K_{IJ} \bar{r}_J.$$

Now look at $\mathbf{d}_g^t \mathbf{k}_g \bar{\mathbf{d}}_g$ for an element. The element local displacements are simply values of $\mathbf{r}$ with different labels. So if d_i and d_j are the same as r_I and r_J then the work term

$$d_i k_{ij} \bar{d}_j$$

must be the same as the previous one. In other words k_{ij} is simply inserted in $\mathbf{K}$ at the position (I, J). Of course (2.14) is a summation and therefore there will be more than one contribution to K_{IJ} from adjoining elements sharing the same r_I and r_J. The global stiffness assembly is therefore simply a book-keeping exercise whereby all the element stiffness matrices have their components $[i, j]$ identified with the global list $[I, J]$, and the total K_{IJ} is summed automatically as a routine process. The selection matrix $\mathbf{a}_g$ in the product (2.16) is therefore simply identifying local (i, j) with global (I, J) and this expensive product is never formed. We will continue to use the expression simply for convenience since otherwise we would have to write $\mathbf{K} = \sum \mathbf{k}$ which is also symbolic and does not convey the nature of the summation.

It is essential that the assembly process should be routine in any well-ordered finite element system, whatever the nature of the elements. Special features such as supports are best left to the end after $\mathbf{K}$ has been assembled. For example if the Ith displacement at a support is zero then all of the Ith *column* in $\mathbf{K}$ will be multiplied by zero in $\mathbf{R} = \mathbf{K}\mathbf{r}$. The Ith *row* of $\mathbf{K}$ could still be multiplied by all the other displacements to produce a force R_I which would however be a support reaction rather than a known applied load. We therefore simply remove both row and column associated with a zero displacement, a very routine procedure at the end of the assembly. The reader can confirm this routine nature by solving the example of Figure 2.6a. We now return to the element stiffness $\mathbf{k}$ which came out of the work equation.

2.4 Framework and Element Stiffnesses

Consider the bar element virtual work $\mathbf{d}_g^t \mathbf{k}_g \bar{\mathbf{d}}_g$ again in (2.14). Since this is virtual work we know that the first product must be element 'forces', say $\mathbf{P}_g$, doing work $\mathbf{P}_g^t \bar{\mathbf{d}}_g$, and corresponding to the displacements $\mathbf{d}_g$. If $\mathbf{P}_g = \mathbf{k}_g \mathbf{d}_g$ it is natural to denote $\mathbf{k}_g$ as an element stiffness – which is why we chose the label. These element forces $\mathbf{P}_g$ did not need a physical interpretation, and neither did $\mathbf{k}_g$ for that matter, they simply appeared in the summation process. In fact in later two- and three-dimensional elements the physical concept of concentrated element forces can be confusing since a *real* concentrated force in a thin plate would lead to infinite stresses! However in a bar element they do make sense if we consider an isolated bar, in which $\mathbf{r}_g$ and $\mathbf{d}_g$ are the same four components, and (2.14) becomes

$$\mathbf{R}^t \bar{\mathbf{r}} = \mathbf{P}_g^t \bar{\mathbf{d}}_g = N \bar{\Delta} = N \mathbf{B} \bar{\mathbf{d}}_g.$$

Therefore

$$\mathbf{P}_g = \mathbf{B}^t N = \begin{bmatrix} -N\cos\theta \\ -N\sin\theta \\ N\cos\theta \\ N\sin\theta \end{bmatrix}.$$

This set of four forces (see Figure 2.4) is clearly in equilibrium with the bar force N.

If we turn to the element stiffness (2.15) and use (2.13) we find

$$\mathbf{k}_g = \mathbf{B}^t k \mathbf{B} = \frac{AE}{L} \begin{bmatrix} \cos^2\theta & \sin\theta\cos\theta & -\cos^2\theta & -\sin\theta\cos\theta \\ & \sin^2\theta & -\sin\theta\cos\theta & -\sin^2\theta \\ & & \cos^2\theta & \sin\theta\cos\theta \\ \text{symmetric} & & & \sin^2\theta \end{bmatrix} \quad (2.18)$$

Now a single unsupported bar is not a respectable structure. It can be moved in space as a rigid body without straining it and hence with no internal force N or set of forces $\mathbf{P}_g$. This means that there exists a set of rigid body displacements $\mathbf{d}_g$ for which

$$\mathbf{P}_g = \mathbf{k}_g \mathbf{d}_g = \mathbf{0}. \quad (2.19)$$

It is known that this can only be true if the determinant of $\mathbf{k}_g$ vanishes, a fact which the reader can verify using (2.18). The matrix $\mathbf{k}_g$ is said to be *singular*: an appropriate term in view of the form of the inverse (1.11). The zero determinant implies that there is at least one linear relationship between the equations (2.19) so perhaps one displacement could be set to

A FINITE ELEMENT PRIMER

zero and one equation removed, leaving a respectable set. Because there are three possible and independent forms of rigid body motion – shown in Figure 2.5a – it turns out that three displacements have to be suppressed and three equations have to be removed before we achieve a structure like Figure 2.5b. The reader can verify that the determinants of the 3×3, and 2×2 reduced sets are still zero. The *rank* of a matrix is the size of the largest sub-matrix with a non-zero determinant and in this case the stiffness matrix has rank 1 (or its rank 'deficiency' is 3).

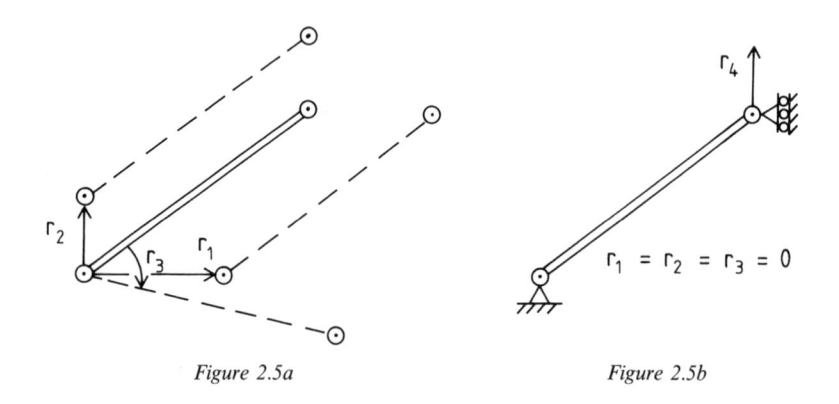

Figure 2.5a   Figure 2.5b

The structure of Figure 2.5b can resist a load R_4 and the stiffness, after deleting the first three rows and columns of (2.18) is simply

$$k_{44} = \frac{AE}{L}\sin^2\theta.$$

Although the isolated bar stiffness matrix $\mathbf{k}_g$ is quite properly singular, it is crucial that the assembled global stiffness $\mathbf{K}$ is not so. The framework should be supported adequately and there should be no internal mechanism. A simple count of bars is not adequate, they have to be in the right places. The reader is invited to form the values of $\mathbf{k}_g$ for $g = 1, 2,$ and 3 in the framework of Figure 2.6a, and then sum all stiffness elements k_{ij} into the global stiffness:

$$\mathbf{K} = \frac{AE}{L}\begin{bmatrix} \frac{1}{2} & 0 & -\frac{1}{4} \\ & \frac{3}{2} & \frac{\sqrt{3}}{4} \\ \text{symmetric} & & \frac{5}{4} \end{bmatrix}.$$

20

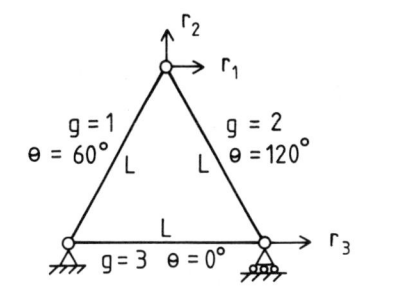

Figure 2.6a

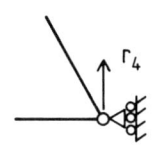

Figure 2.6b

If the formulation of **K** has proved instructive for this respectable structure, then repeat the exercise but this time let the vertical freedom r_4 be unconstrained and the horizontal one zero, at the right hand support, as shown in Figure 2.6b. It should be obvious that this second framework is a mechanism, and the reader will find that its stiffness matrix is singular – just like the unconstrained bar of equation (2.19).

3. General Continuum

3.1 Fundamentals again

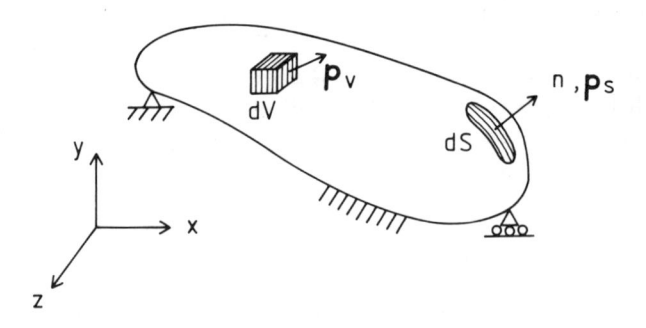

Figure 3.1

Figure 3.1 shows an arbitrarily shaped solid supported in some way. Any point inside the solid is denoted as (x, y, z) and any point on the surface has a local outward-pointing normal 'n' whose orientation is usually described by its three direction cosines $\partial n/\partial x$, $\partial n/\partial y$, and $\partial n/\partial z$. In general there may be two sets of applied forces. Firstly there may be internal body forces inside the volume V whose magnitude per unit volume is denoted by components p_{vx}, p_{vy}, and p_{vz}. For example, a gravity force would be $p_{vy} = -\rho$ (the density) or there might be other inertia forces like centrifugal and so on. It is convenient to write these three components as a single body force vector

$$\mathbf{p}_v = \begin{bmatrix} p_{vx} \\ p_{vy} \\ p_{vz} \end{bmatrix}. \tag{3.1}$$

Likewise there could be tractions per unit area (not necessarily normal pressures) on the surface S, also having three components:

$$\mathbf{p}_s = \begin{bmatrix} p_{sx} \\ p_{sy} \\ p_{sz} \end{bmatrix}. \quad (3.2)$$

Any system of loads has to fall into categories (3.1) or (3.2).

The stresses inside V will have two types of component, the direct stress components σ_{xx}, σ_{yy} and σ_{zz}, and the shear stresses σ_{xy}, σ_{yz}, and σ_{zx}. Now stress is not strictly a vector, like 'force' or 'displacement' having both magnitude and direction. A stress component, say σ_{xy}, will possess magnitude, direction (x), and a plane (normal, y) on which it acts. Nevertheless it is very convenient to represent both stress and strain components as single column matrices. Thus

$$\boldsymbol{\sigma} = \begin{bmatrix} \sigma_{xx} \\ \sigma_{yy} \\ \sigma_{zz} \\ \sigma_{xy} \\ \sigma_{yz} \\ \sigma_{zx} \end{bmatrix} \quad \text{and} \quad \boldsymbol{\varepsilon} = \begin{bmatrix} \varepsilon_{xx} \\ \varepsilon_{yy} \\ \varepsilon_{zz} \\ \varepsilon_{xy} \\ \varepsilon_{yz} \\ \varepsilon_{zx} \end{bmatrix}. \quad (3.3)$$

Turning firstly to the equations of equilibrium inside V; by looking at the small linear increments in stress across an infinitesimal element, $dx\,dy\,dz = dV$, we can equilibrate these changes to applied forces. For example in the 'x' direction

$$\frac{\partial \sigma_{xx}}{\partial x} + \frac{\partial \sigma_{xy}}{\partial y} + \frac{\partial \sigma_{zx}}{\partial z} + p_{vx} = 0 \quad (3.4)$$

and similar equations in the y and z directions. These three equilibrium equations can be very conveniently condensed into a single matrix form

$$\partial^t \boldsymbol{\sigma} + \mathbf{p}_v = 0 \quad (3.5)$$

where the 3 × 6 array of operators ∂ is given by

$$\partial = \begin{bmatrix} \dfrac{\partial}{\partial x} & 0 & 0 \\ 0 & \dfrac{\partial}{\partial y} & 0 \\ 0 & 0 & \dfrac{\partial}{\partial z} \\ \dfrac{\partial}{\partial y} & \dfrac{\partial}{\partial x} & 0 \\ 0 & \dfrac{\partial}{\partial z} & \dfrac{\partial}{\partial y} \\ \dfrac{\partial}{\partial z} & 0 & \dfrac{\partial}{\partial x} \end{bmatrix}. \qquad (3.6)$$

A typical infinitesimal element will look different at the surface where there are surface tractions on the inclined face. It is found that the three equations of equilibrium on the surface can be summarised as

$$[\partial^t n]\sigma = \mathbf{p}_s \qquad (3.7)$$

where $\partial^t n$ implies that all the operators in ∂^t are acting upon 'n' to form $\partial n/\partial x$, $\partial n/\partial y$ etc. which are the direction cosines of the normal. Thus (3.7) ensures that the correct components are taken.

Secondly the equations of compatibility, relating strains to displacement components u, v, w, are the familiar forms

$$\varepsilon_{xx} = \frac{\partial u}{\partial x},\ \varepsilon_{xy} = \frac{\partial u}{\partial y} + \frac{\partial v}{\partial x},\ \text{etc.}$$

It is natural to form the single displacement vector

$$\mathbf{u} = \begin{bmatrix} u \\ v \\ w \end{bmatrix}. \qquad (3.8)$$

Then all six compatibility equations can be summarised as

$$\varepsilon = \partial \mathbf{u}. \qquad (3.9)$$

(The inquisitive reader who wishes to know why the same operator ∂ appears conveniently in (3.5), (3.7) and (3.9), should consult ref. 2 or the end of the next section, equations (3.12) to (3.15).)

Finally there will be a material stress-strain law relating all six strains to all stress components. For the purposes of this Primer it will be assumed that this law is linear (Hookean) involving 21 material constants. In the case of an *isotropic* material these reduce to two in number, Young's Modulus E, and Poisson's ratio v; that is

$$\varepsilon = \mathbf{f}\sigma$$

where

$$\mathbf{f} = \frac{1}{E} \begin{bmatrix} 1 & -v & -v & 0 & 0 & 0 \\ & 1 & -v & 0 & 0 & 0 \\ & & 1 & 0 & 0 & 0 \\ & & & 2(1+v) & 0 & 0 \\ & \text{symmetric} & & & 2(1+v) & 0 \\ & & & & & 2(1+v) \end{bmatrix}. \quad (3.10)$$

In the presence of thermal strains we would simply add the separate effects of stress and temperature

$$\varepsilon = \mathbf{f}\sigma + \mathbf{\eta}$$

where for an isotropic material, coefficient of expansion α,

$$\mathbf{\eta}^t = [\alpha T \quad \alpha T \quad \alpha T \quad 0 \quad 0 \quad 0].$$

The inverse of $\mathbf{f}$ is the material stiffness $\mathbf{E}$, given by

$$\mathbf{E} = \frac{E}{(1-2v)(1+v)} \begin{bmatrix} 1-v & v & v & 0 & 0 & 0 \\ & 1-v & v & 0 & 0 & 0 \\ & & 1-v & 0 & 0 & 0 \\ & & & \tfrac{1}{2}-v & 0 & 0 \\ & \text{symmetric} & & & \tfrac{1}{2}-v & 0 \\ & & & & & \tfrac{1}{2}-v \end{bmatrix} \quad (3.11)$$

Summarising Figure 2.2 once more

A FINITE ELEMENT PRIMER

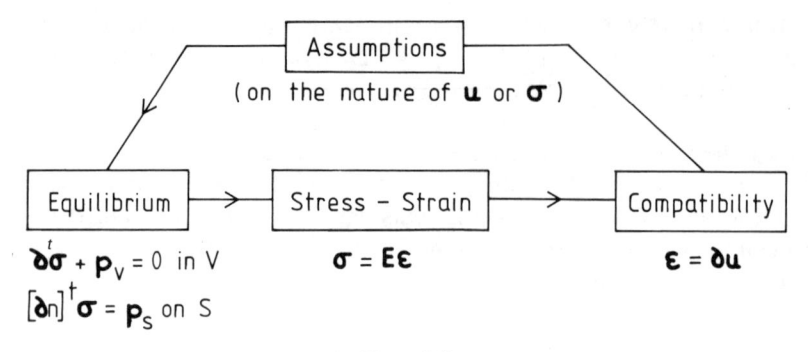

Figure 3.2

Equilibrium: $\partial^t\sigma + p_v = 0$ in V; $[\partial n]^t \sigma = p_s$ on S

Stress – Strain: $\sigma = E\varepsilon$

Compatibility: $\varepsilon = \partial u$

Assumptions (on the nature of **u** or **σ**)

The next step is to establish a general PVD and show that it delivers the equilibrium equations.

3.2 PVD for any solid

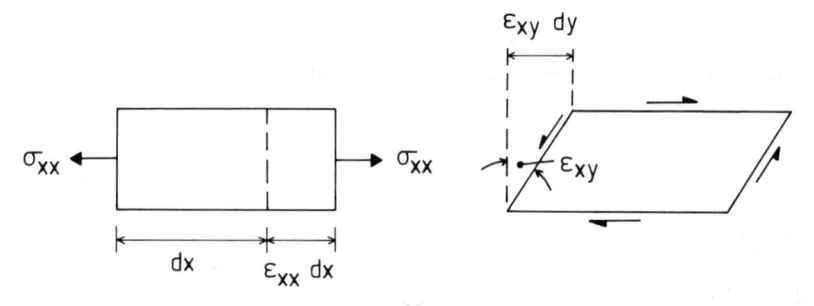

Figure 3.3

Two sorts of deformation cause stresses to do work, extension and shear, as shown in Figure 3.3. The work done in each case is $\sigma_{xx}\varepsilon_{xx}$ and $\sigma_{xy}\varepsilon_{xy}$, per unit volume. The total internal virtual work is therefore

$$\int_V (\sigma_{xx}\bar{\varepsilon}_{xx} + \sigma_{yy}\bar{\varepsilon}_{yy} + \sigma_{zz}\bar{\varepsilon}_{zz} + \sigma_{xy}\bar{\varepsilon}_{xy} + \sigma_{yz}\bar{\varepsilon}_{yz} + \sigma_{zx}\bar{\varepsilon}_{zx}) dV = \int_V \sigma^t \bar{\varepsilon} dV.$$

The work done by both body forces and surface forces over displacements u is simply

$$\int_V p_v^t \bar{u} \, dV \quad \text{and} \quad \int_S p_s^t \bar{u} \, ds.$$

GENERAL CONTINUUM

As before, the PVD simply equates both work quantities:

$$\boxed{\int_V \sigma^t \bar{\varepsilon}\, dV = \int_V \mathbf{p}_v^t \bar{\mathbf{u}}\, dV + \int_S \mathbf{p}_s^t \bar{\mathbf{u}}\, ds.} \quad (3.12)$$

This very important fundamental work equation is the basis of the finite element displacement method. It will enable us to extract all the necessary finite element equations, properties, and so on for any structure. The process will turn out to be quite automatic once the assumptions have been made for the type of displacement field **u** existing in the structure. Before looking at several examples and consequences we should therefore show that equation (3.12) will deliver. As it stands it delivers nothing, but we have learned to enforce compatibility equations (3.9), so inserting $\varepsilon = \partial \mathbf{u}$,

$$\int_V \sigma^t\, \partial \bar{\mathbf{u}}\, dV - \int_V \mathbf{p}_v^t \bar{\mathbf{u}}\, dV - \int_S \mathbf{p}_s^t \bar{\mathbf{u}}\, ds = 0. \quad (3.13)$$

Unlike the framework examples we are not immediately able to say that $\bar{\mathbf{u}}(x, y, z)$ is arbitrary everywhere and its coefficient is therefore zero, since (3.13) has $\bar{\mathbf{u}}$ in two forms. However it is possible to convert $\partial \mathbf{u}$ to $\mathbf{u}$ using a three dimensional matrix form of 'integration by parts'. This says that

$$\int_V (\mathbf{u}^t\, \partial^t \boldsymbol{\sigma} + \boldsymbol{\sigma}^t\, \partial \mathbf{u})\, dV = \int_S \mathbf{u}^t [\partial^t n] \boldsymbol{\sigma}\, ds. \quad (3.14)$$

The important point about this theorem (due to Gauss) is that it is only true provided $\partial^t \boldsymbol{\sigma}$ and $\partial \mathbf{u}$ are finite in V, that is *the stress and displacement fields are continuous*. We return to this important point when discussing finite element approximations.

On substituting (3.14) into (3.13) we obtain the desired consequence of the PVD

$$\int_V \bar{\mathbf{u}}^t [\partial^t \boldsymbol{\sigma} + \mathbf{p}_v]\, dV - \int_S \bar{\mathbf{u}}^t [(\partial^t n)\boldsymbol{\sigma} - \mathbf{p}_s]\, ds = 0. \quad (3.15)$$

By stipulating that $\bar{\mathbf{u}}(x, y, z)$ is an arbitrary but continuous function everywhere, its coefficients must vanish, so once again the equations of equilibrium emerge from the square brackets in (3.15):

$$\partial^t \boldsymbol{\sigma} + \mathbf{p}_v = \mathbf{0} \quad \text{in } V \quad (3.5)$$

and

$$(\partial^t n)\sigma = \mathbf{p}_s \quad \text{on } S. \tag{3.7}$$

It is important to notice that our fundamental theorem has supplied equilibrium conditions not only inside the solid but also on the surface if required. When we come to use approximate finite element techniques it is not necessary to worry about the equilibrium boundary conditions. The PVD will do the worrying. Part of the surface S will of course be held and supported in some way, and here the tractions $\mathbf{p}_s$ will be unknown reactions and not specified loads. Although equations (3.12) and (3.15) are still valid here, it is conventional to remove the unknown reactions by simply choosing (as we are free to do) the virtual displacements $\bar{\mathbf{u}}$ to be zero over such parts of the surface.

The power of PVD (3.12) and the importance of (3.15) become obvious only when we choose finite element approximations, but first let us look at simple beam problems, exact and approximate. Simple beams are familiar, and they have continuously varying displacements (which we avoided in pinjointed frameworks.)

4. Beams

4.1 Exact Solutions

Beams and rigid-jointed frameworks are some of the most common forms of structure used in the civil engineering and building industries, so it is appropriate to examine them in their own right. Their popularity, and their relative analytical simplicity, has led to many special-purpose finite element type programs being developed, particularly of late for microcomputers. We will adopt the PVD strategy as promised:

a. Make all necessary assumptions for $\mathbf{u}(x, y, z)$.
b. Enforce compatibility, $\varepsilon = \partial \mathbf{u}$.
c. Apply the PVD and carry out integrations if possible.
d. Extract equations of equilibrium from coefficients of $\bar{\mathbf{u}}$ and solve for $\mathbf{u}$.

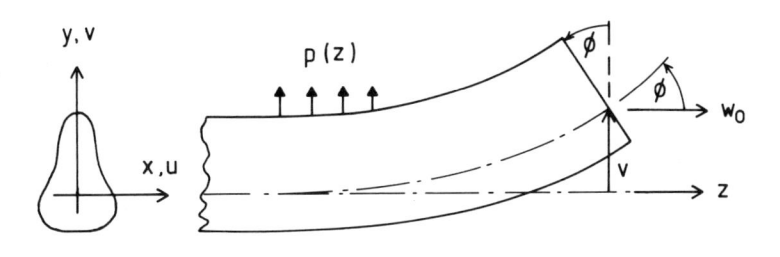

Figure 4.1

(*a*) *Assumptions*

For simplicity we take the beam cross-section to be symmetrical about the y axes, and the beam to be loaded only in this direction ($p(z)$ per unit length). This will ensure no horizontal transverse deflections, u.

A FINITE ELEMENT PRIMER

The cross section is assumed to be compact, and no distortion in the x-y plane considered. The vertical deflection, v, is therefore a function of z only.

The z axis will be taken as the centroidal axis for which

$$\int_A y \cdot dA = 0$$

and the axial displacement of this denoted as $w_0(z)$.

Plane sections normal to the centroidal axis are assumed to remain plane and normal to the deflected axis; so a small rotation ϕ of the axis is the same as that of the section (see Figure 4.1). The total axial displacement of any point on the section is therefore

$$w = w_0 - y\phi = w_0 - yv'(z) \tag{4.1}$$

abbreviating $\partial v/\partial z$ by a prime.

All assumptions have been made and the complete displacement field is

$$\mathbf{u} = \begin{bmatrix} u \\ v \\ w \end{bmatrix} = \begin{bmatrix} 0 \\ v(z) \\ w_0(z) - yv'(z) \end{bmatrix}$$

(b) *Compatibility*

Inserting the above into $\varepsilon = \partial u$ (3.9) we obtain using (3.6)

$$\varepsilon = \begin{bmatrix} \varepsilon_{xx} \\ \varepsilon_{yy} \\ \varepsilon_{zz} \\ \varepsilon_{xy} \\ \varepsilon_{yz} \\ \varepsilon_{zx} \end{bmatrix} = \begin{bmatrix} 0 \\ 0 \\ w'_0 - yv'' \\ 0 \\ 0 \\ 0 \end{bmatrix}$$

that is the only survivor is the component ε_{zz}.

(c) *PVD*

Equation (3.12) becomes, using $\sigma_{zz} = E\varepsilon_{zz}$, and taking the beam to be of length L,

$$\int_0^L \int_A E(w_0' - yv'')(\bar{w}_0' - y\bar{v}'') dA\, dz = \int_0^L p(z)\, dz$$

or

$$\int_0^L (EA w_0' \bar{w}_0' + EI v'' \bar{v}'' - p\bar{v})\, dz = 0 \tag{4.2}$$

where

$$I = \int_A y^2\, dA \quad \text{and} \quad \int y\, dA = 0.$$

This simple result is possible because we have been able to integrate over the cross section, and it explains the attractions of beam theory. We can in fact now go on and extract the full equations by converting $\bar{w}_0'$ to $\bar{w}_0$ and $\bar{v}''$ to $\bar{v}$ by integrating by parts. Since $\bar{v}$ and $\bar{w}_0$ are unrelated they may be examined separately, and in this problem there is no cross coupling with the real displacements v and w. The $\bar{w}_0(z)$ terms produce

$$-\int_0^L (EA w_0')' \bar{w}_0\, dz + [EA w_0' \bar{w}_0]_0^L = 0.$$

From the integral, $(EA w_0')'$ is zero, which is a simple equilibrium statement that the mean axial load does not change. If A is constant, then $w_0(z)$ varies linearly, and if we further assume that the beam is held axially at the ends then $w_0(z)$ is zero. The beam has consequently become a pure bending problem for which the remainder of (4.2) becomes

$$\int_0^L [(EI v'')'' - p] \bar{v}\, dz + [(EI v'') \bar{v}' - (EI v'')' \bar{v}]_0^L = 0. \tag{4.3}$$

The coefficient of $\bar{v}$ in the integrand then delivers

$$(EI v'')'' = p(z). \tag{4.4}$$

This is a fourth order differential equation (of equilibrium) and consequently four boundary conditions are required, two at each end. If the boundary conditions are *kinematic* like 'no displacement' ($v = 0$) or 'fully clamped' ($v' = 0$ also) then these are enforced directly. As we

A FINITE ELEMENT PRIMER

mentioned, at such supports v and v' are put to zero. The whole of (4.3) is now zero. But the beam end may not be held, in which case the coefficients of v or v' in square brackets in (4.3) will deliver

$$\bar{v} \neq 0 \rightarrow (EIv'')' = 0 \quad \text{i.e. no shear}$$

$$\bar{v}' \neq 0 \rightarrow (EIv'') = 0 \quad \text{i.e. no moment.}$$

Now this primer is concerned with the finite element method and so we now assume that (4.4) cannot be solved in closed form. This assumption would of course be entirely justified if the rigidity EI was an inconvenient function of z (like an aircraft wing).

4.2 Approximate Finite Element Solution

It may be difficult to solve differential equations but it should be easier to *integrate* known functions even if we have to do this in an approximate fashion. This is the heart of the finite element method when applied to structures over which the displacements vary in some continuous but unknown fashion. The first step is to divide the beam into smallish (finite) elements of length l which are made small enough so that variations in $v(z)$, $EI(z)$, and $p(z)$ may be approximated, even if the magnitudes of $v(z)$ remain to be found. Recalling that an integral is merely a summation (over infinitesimal elements) the PVD (4.2) becomes

$$\sum \left[\int_0^l EIv''\bar{v}'' \, dz \right] = \sum \left[\int_0^l p\bar{v} \, dz \right] \tag{4.5}$$

where the summations on both sides of the equation are carried out over all elements.

Thus the finite element method may be viewed simply as a strategy for evaluating an integral numerically. The left hand side of (4.5) is a single scalar quantity (work) which varies over the length of the beam, and we will be attempting to obtain the area under this 'work length' curve in terms of discrete values at selected points along the length (cf. Simpson's Rule). Later we even extend this idea to the element integrals themselves when they become too cumbersome to evaluate in closed form.

We now have to guess the form of $v(z)$ over $0 < z < 1$, to be able to perform the integral. Also the elements should fit together so it is necessary to insist that $v(z)$ is continuous from one element to the next and equally so for the rotation $v'(z)$ – otherwise a fictitious hinge would be implied. The trick then is to use these quantities as the unknown magnitudes at the interface of one element and the next. The ends of the elements are referred to as '*nodes*' and the unknown displacements and slopes at these ends as '*nodal values*', $v(0)$, $v'(0)$, $v(1)$, and $v'(1)$. This concept is later generalised in two and three dimensions.

The element displacement is therefore written as, for $0 < z < 1$,

$$v(z) = v(0)N_1(z) + v'(0)N_2(z) + v(1)N_3(z) + v'(1)N_4(z)$$

where the $N(z)$ are assumed 'shape functions' or 'interpolation functions'. Since this expansion has to produce the required values of $v(z)$ and $v'(z)$ at both ends, whatever the four nodal values are, then it is necessary that $N_1(z)$ should have unit value at $z=0$, zero value at $z=1$ and zero values of gradient at both $z=0$ and 1. Similar specifications for $N_2(z)$, $N_3(z)$, and $N_4(z)$ will ensure that the above expansion produces correct values at the nodes. A set of shape functions with these characteristics are the 'Hermitian' cubic polynomials shown in Figure 4.2. We have changed the

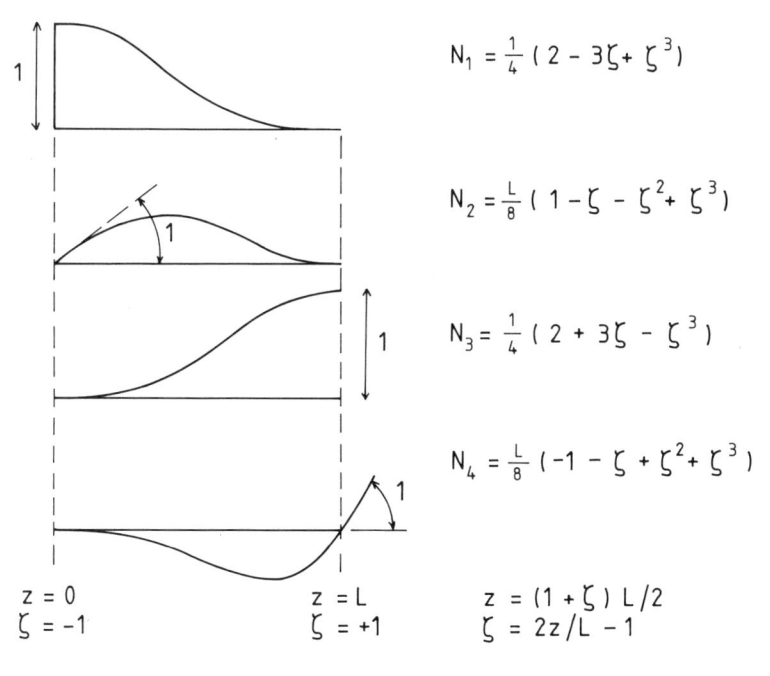

$$N_1 = \tfrac{1}{4}(2 - 3\zeta + \zeta^3)$$

$$N_2 = \tfrac{L}{8}(1 - \zeta - \zeta^2 + \zeta^3)$$

$$N_3 = \tfrac{1}{4}(2 + 3\zeta - \zeta^3)$$

$$N_4 = \tfrac{L}{8}(-1 - \zeta + \zeta^2 + \zeta^3)$$

$z = 0$ $z = L$ $z = (1 + \zeta)L/2$
$\zeta = -1$ $\zeta = +1$ $\zeta = 2z/L - 1$

Figure 4.2

variable from z to a more convenient non-dimension ζ to simplify later integrations.

These cubic shapes are *assumed* and may not represent an exact solution. If fact if we examine the governing differential equation (4.4) we see that the deflected shape will be a cubic only if the element has uniform stiffness EI and is not loaded between nodes – that is $p(z) = 0$. This statement is true of all shape functions, N, whether the element is a beam, plate, brick and so on. *They are approximate*, and we hope that over a reasonably small element the shape functions do not differ too much from the real ones. In general then our solution will be approximate, so that $(EIv'')'' \neq p(z)$. There will therefore be an error; a *residual* out-of-balance force between the applied force $p(z)$ and the resistance $(EIv'')''$. If we turn to the PVD in the form (4.3) it is apparent that our technique ensures that this residual force does no work over the virtual displacement $\bar{v}(z)$. However $\bar{v}(z)$ is no longer a completely arbitrary function at every point along the beam since its shape has been assumed. The aim then is to introduce as many variations into $v(z)$ as possible so that by enforcing the integrals to be zero for all virtual degrees of freedom, there is a sporting chance that the residual will be fairly small everywhere. The number of variables in $v(z)$ is dictated by the degrees of freedom of the element (four in this case) and the number of elements, that is the smaller the element the better.

The residual error may be physically imagined as a continuous artificial loading (or constraint) forcing the beam to maintain the approximate deflection. If we were to gradually remove this constraint, and allow the beam to move to its true position of equilibrium, then positive work must be done by $p(z)$. If this were not so then the beam would not move to the true position on releasing the constraints, but find a better least work alternative. Thus the elastic potential energy of the beam must increase as the constraints are decreased. We must therefore expect *the strain energy in a structure, when found in terms of an approximate solution using the PVD, to be less than the true strain energy*. Moreover as we increase the number of degrees of freedom in $v(z)$ we would expect the solved strain energy to increase gradually and approach the true value. The solution is consequently referred to as a *lower bound*, or a solution in which the structure is *over-stiff*. Remember though that this is a bound 'in the mean' on total strain energy and not on the stress or displacement at a point. Local stresses may be higher than the true ones. (There are methods for finding true bounds at a point but as yet they are not considered commercially viable, and it is still conventional to increase the number of elements to test whether the solved solutions are approaching the true one. Actually further solutions cost money, and it is even more conventional to look solely at residual errors, wherever they are examinable, and decide that the answer is good enough.)

Consider now the remaining terms in square brackets in (4.3). They will involve nodal values of $\bar{v}$ and $\bar{v}'$ whose coefficients must also vanish. Normally at node 'i' say there will be two contributions from adjacent elements to the left (L) and to the right (R) of node 'i' forming

$$[(EIv'')'_R - (EIv'')'_L]\bar{v}_i = 0$$

so the PVD would ensure that the shear $(EIv'')'$ is continuous through the node, provided there was not a concentrated applied force there. But we shall find that the pressure loading integral, $\int p\bar{v}\,dz$ contains terms in $\bar{v}_i$ which appear as a consequence of the finite element approximation, and this therefore inputs a discontinuity into the shear. A similar input will occur for the bending moment EIv''. Thus this particular finite element may produce discontinuous moments and stresses at nodes if there is loading between nodes. This in itself is not a bad thing since it is an indication of whether the idealisation is adequate for the loading. Smaller elements may be required. These points deserve further consideration when discussing two and three-dimensional elements.

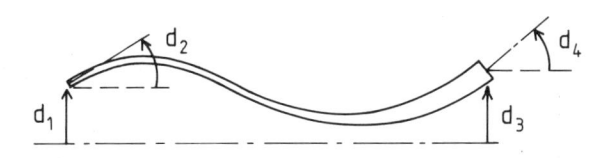

Figure 4.3

It is now worth considering in detail the element integrals in (4.5). For convenience write the shape functions of Figure 4.2 as a single (1 × 4) row matrix

$$\mathbf{N} = [N_1, N_2, N_3, N_4]. \qquad (4.6)$$

Denote the nodal displacements again by $\mathbf{d}_g$ (see Figure 4.3) so that

$$v(z) = [N_1 d_1 + N_2 d_2 + N_3 d_3 + N_4 d_4] = \mathbf{N}\mathbf{d}_g. \qquad (4.7)$$

A FINITE ELEMENT PRIMER

The internal virtual work of the element is therefore

$$\int_0^l EI v'' \bar{v}'' \, dz = \int_{-1}^1 \mathbf{d}_g^t (\mathbf{N}'')^t EI \mathbf{N}'' \bar{\mathbf{d}}_g (8/l^3) \, d\zeta = \mathbf{d}_g^t \mathbf{k}_g \bar{\mathbf{d}}_g$$

(cf. frameworks)

where this time the element stiffness matrix (if $EI = $ constant)

$$\mathbf{k}_g = \frac{8EI}{l^3} \int_{-1}^1 (\mathbf{N}'')^t \mathbf{N}'' \, d\zeta$$

or

$$\mathbf{k}_g = \frac{2EI}{l^3} \begin{bmatrix} 6 & 3l & -6 & 3l \\ & 2l^2 & -3l & l^2 \\ \text{symmetric} & & 6 & -3l \\ & & & 2l^2 \end{bmatrix}. \quad (4.8)$$

Turning to the right-hand side of (4.5), the external virtual work, we have:

$$\int_0^l p\bar{v} \, dz = \int_{-1}^1 p(\zeta) \mathbf{N}^t \bar{\mathbf{d}}_g (l/2) \, d\zeta.$$

This work can obviously be written again as element 'forces' $\times$ displacements, or $\mathbf{P}_g^t \bar{\mathbf{d}}_g$ where

$$\mathbf{P}_g = \frac{l}{2} \int_{-1}^1 \mathbf{N} p(\zeta) \, d\zeta. \quad (4.9)$$

These 'forces' $\mathbf{P}_g$ correspond to the displacements $\mathbf{d}_g$ so that, noting Figure 4.3, P_1 is a shear force and P_2 a moment. They are called *'kinematically equivalent' nodal forces* since they replace a distributed force $p(\zeta)$ weighted with the shape functions $\mathbf{N}$ so that the correct work is simulated. To replace $p(\zeta)$ by statically equivalent forces would be incorrect and would introduce errors in the solved displacements. For example if $p(\zeta)$ is a constant loading p, equation (4.9) produces

$$\mathbf{P}_g = \begin{bmatrix} \dfrac{pl}{2} \\[6pt] \dfrac{pl^2}{12} \\[6pt] \dfrac{pl}{2} \\[6pt] \dfrac{-pl^2}{12} \end{bmatrix}.$$

BEAMS

$$\frac{pl^2}{12} \qquad \frac{pl^2}{12}$$

$$\frac{pl}{2} \qquad \frac{pl}{2} \qquad\qquad \frac{pl}{2} \qquad \frac{pl}{2}$$

kinematically equivalent loads   incorrect statically equivalent loads

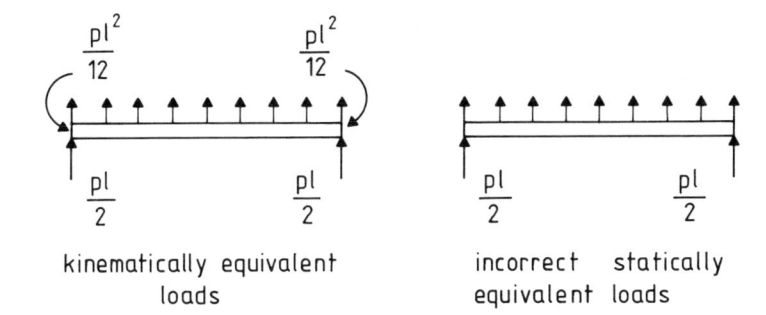

Figure 4.4

(It can be shown that the kinematically equivalent loads are those which, if applied in the opposite sense as constraints, would keep all nodal displacements zero in the presence of the true loading.)

Having found both $\mathbf{k}_g$ and $\mathbf{P}_g$, inserting the kinematic connectivity $\mathbf{d}_g = \mathbf{a}_g \mathbf{r}$ into (4.5), the summation can be performed and the left-hand side produces the global stiffness again as

$$\mathbf{K} = \sum \mathbf{a}_g^t \mathbf{k}_g \mathbf{a}_g. \qquad (2.16)$$

The right-hand side becomes

$$\sum_g \mathbf{P}_g^t \mathbf{d}_g = (\sum \mathbf{P}_g^t \mathbf{a}_g)\bar{\mathbf{r}} = \mathbf{R}^t \bar{\mathbf{r}}.$$

So the global loading vector is

$$\mathbf{R} = \sum_g \mathbf{a}_g^t \mathbf{P}_g. \qquad (4.10)$$

This last statement simply means that all kinematically equivalent nodal forces P_i arising from elements which share a common displacement r_i are summed as R_i.

Having solved $\mathbf{Kr} = \mathbf{R}$ it is routine to backtrack and recover the gth element strains. Thus $\mathbf{d}_g = \mathbf{a}_g \mathbf{r}$, then $\varepsilon = -yv'' = -y\mathbf{N}''\mathbf{d}$.

Two simple three-element solutions are shown in Figure 4.5 to illustrate typical behaviours. In both cases the finite element displacements are not

A FINITE ELEMENT PRIMER

shown since they are indistinguishable from the correct values, but the stresses are not so accurate. Strains are the derivatives of an approximate displacement (in beams the *second* derivatives) and differentiation inevitably decreases accuracy.

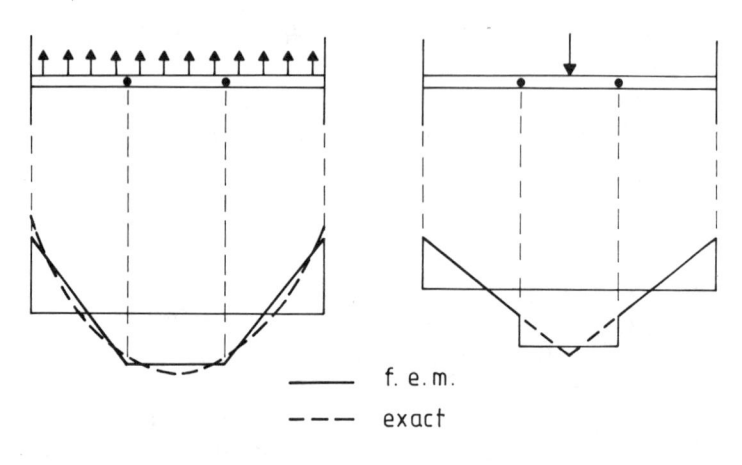

――― f. e. m.
― ― ― exact

Figure 4.5a   *Figure 4.5b*

The element's linear variation in moment can hardly do better in approximating the true parabola in Figure 4.5a. No nodal discontinuities are present because the kinematically equivalent nodal moments cancel. This is not true in Figure 4.5b where applied nodal moments guarantee a 'jump' in stresses. Of course a good engineer would place an element node under the concentrated force – and get the exact answers.

If we were to increase the number of elements and decrease their size, the errors would decrease. This feature is illustrated in Section 5. However two features should be noted at this stage, which are true of all finite element models, and which will guarantee convergence as element sizes are reduced.

i. Although equilibrium may not be satisfied exactly at every interior point or across interfaces, an element as a whole should be in equilibrium. The *structure* as a whole should be in equilibrium, but it will be if each element is. Now the *resultant* forces and moments on an element correspond to virtual rigid body displacements and rotations, so one criterion should be:

An element should describe rigid body modes exactly.

In the case of our beam element there are only two such modes, a vertical translation and a rotation. Taking the first as $\mathbf{d}_g^t = [1, 0, 1, 0]$ and the second as $\mathbf{d}_g^t = [0, 1, l, 1]$ then using $v(\zeta) = \mathbf{Nd}$ from (4.7) confirms that $v(\zeta)$ is constant and linear respectively.

ii. It can be shown (ref. 1) that *an element should be capable of simulating constant strain states*, which must be the ultimate destiny as element sizes shrink to zero.

In the case of our beam element, the bending moments vary linearly so the constant value is indeed represented as a special case. This can be confirmed by putting $\mathbf{d}_g = [0, 1, 0, -1]$ whence

$$v = N_2 - N_3 = \frac{l}{4}(1 - \zeta^2)$$

so $v'' = $ constant.

4.3 Rigid Jointed Frameworks

Most multistorey buildings consist of rigid jointed frameworks with columns and beams of steel or reinforced concrete. The global stiffness for such frameworks can be assembled using the usual procedure (2.16) but the beam element should include longitudinal freedoms since beams (horizontal) and columns (vertical) are to be joined, and all nodal displacement components must match. If the framework is two-dimensional then this simply means adding the reduced axial freedoms of Figure 2.4 (with $\theta = 0$) to Figure 4.3, and producing Figure 4.5.

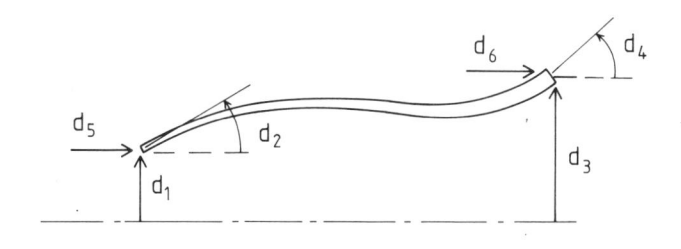

Figure 4.5

There is no coupling between the bending and stretching freedoms in this simple element, so the two stiffness matrices add as follows.

A FINITE ELEMENT PRIMER

$$\mathbf{k}_g \begin{bmatrix} \mathbf{k}_{g\text{(bending)}} & | & \\ ----+----- \\ & | & \mathbf{k}_{g\text{(stretching)}} \end{bmatrix}. \qquad (4.11)$$

The lower right $\mathbf{k}_g$(stretching) stiffness from (2.18), deleting freedoms 2 and 4, with $\theta = 0$, is

$$\mathbf{k}_g(\text{stretching}) = \frac{AE}{l} \begin{bmatrix} 1 & -1 \\ -1 & 1 \end{bmatrix}.$$

We therefore see that the ratio of the bending terms to the stretching terms in (4.11) is of order $(EI/l^3) \div (AE/l)$ which is $(h/l)^2$ where 'h' is the relevant radius of gyration. Now for slender beams this (slenderness) ratio h/l may be as small as $1/20$ or $1/50$, so the stiffness matrix may possibly be ill-conditioned. This is more of a problem with microcomputers of limited wordlength. Large mainframes can usually cope adequately with terms whose significant figures are five orders less than other terms, and it is unusual for users to be concerned about this framework problem. Users of microcomputers should check that the solution procedure can handle such systems of equations. It may be necessary for example to partition $\mathbf{k}$ into bending and stretching sub-matrices and then scale the deflection variables to bring all terms in $\mathbf{K}$ within the same order. Alternatively all beams can be made completely inextensional by imposing displacement constraints which make the two axial displacements in any beam the same, and thereby reducing the number of displacement freedoms.

Most frameworks are of course not two-dimensional and for three-dimensional structures the element nodal freedoms must be increased to three displacements and three rotations at each node.

4.4 Stiffness Transformation

Many rigid jointed frameworks will also include members inclined to the global axes and it is necessary therefore to find the stiffness matrix of an inclined beam, keeping its nodal displacement freedoms aligned to the global axes so that $\mathbf{K}$ can be assembled in the usual way. Instead of using a general expression for element stiffness (like 2.18) it is more convenient to find $\mathbf{k}_g$ using a local set of axes like Figure 4.5 and then transform to any new set that the structure demands.

If the new set of axes (x', y', z') is inclined to the global set (x, y, z) it is first necessary to find the direction cosines $\partial x'/\partial x$, $\partial y'/\partial y$,...etc. We label these as a set

$$T = \begin{bmatrix} \dfrac{\partial x'}{\partial x} & \dfrac{\partial x'}{\partial y} & \dfrac{\partial x'}{\partial z} \\ \dfrac{\partial y'}{\partial x} & \dfrac{\partial y'}{\partial y} & \dfrac{\partial y'}{\partial z} \\ \dfrac{\partial z'}{\partial x} & \dfrac{\partial z'}{\partial y} & \dfrac{\partial z'}{\partial z} \end{bmatrix}. \quad (4.12)$$

It is easily shown that a vector quantity in the old coordinates transforms into the new coordinates as

$$\mathbf{r'} = \mathbf{Tr} \text{ (displacements)} \quad \text{or} \quad \mathbf{P'} = \mathbf{TP} \text{ (forces)}.$$

Now the work done by forces over displacements must be the same, however we take components. Thus

$$\mathbf{P}^t \mathbf{r} = \mathbf{P'}^t \mathbf{r'} = \mathbf{P}^t \mathbf{T}^t \mathbf{Tr}$$

or

$$\mathbf{T}^t \mathbf{T} = \mathbf{I}, \text{ the } 3 \times 3 \text{ unit matrix.}$$

(The properties which emerge from this, using (4.12), turn out to be familiar trigonometric identities for direction cosines.) Equating work again, and using this time the element stiffness $\mathbf{k}$ (in the old) and $\mathbf{k'}$ (the new), we have

$$\mathbf{r'}^t \mathbf{k'} \mathbf{r'} = \mathbf{r}^t \mathbf{k} \mathbf{r} = \mathbf{r}^t \mathbf{T}^t \mathbf{k'} \mathbf{Tr} = \mathbf{r}^t \mathbf{k} \mathbf{r}$$

therefore

$$\mathbf{k} = \mathbf{T}^t \mathbf{k'} \mathbf{T}.$$

Or, pre- and post-multiplying by $\mathbf{T}$ and $\mathbf{T}^t$, and taking the transpose

$$\mathbf{k'} = \mathbf{T}\mathbf{k}\mathbf{T}^t \quad \text{as required.} \quad (4.13)$$

The conversion of element stiffness from local to global measures

A FINITE ELEMENT PRIMER

is a standard routine in most finite element systems (equation (4.13) is known as a congruent transformation). It is useful for beams, rectangular plates and bricks which have simple local descriptions, but less so for the curvilinear shapes discussed in the next section.

5. Two Dimensional Membranes

5.1 Fundamental Assumptions

Two-dimensional 'thinwalled" structures are very common. In all forms of transport, lightness is a virtue, for obvious reasons. It used to be the practice to make a framework and cover it, but the 'stressed skin' is now the norm, and very efficient automobiles, aircraft and rolling stock are the result. Thinwalled pressure vessels are ubiquitous, and box girders have threatened the traditional bridge girders. Ships of all sizes have cellular thinwalled configurations.

These structures are usually designed to be primarily loaded in their plane and to resist loads by membrane action rather than bending. Thin plates are fairly inefficient in bending, unless they can be curved into shells, in which case membrane stresses are able to resist normal loads by virtue of the shell's curvature. Many structures however have to be flat for other reasons and we will look first at two-dimensional flat plates loaded purely in their own $(x\text{-}y)$ plane and having stress components uniform through the plate's thickness – a pure membrane action.

Before looking at specific shapes of membrane elements with three or four sides, both straight and curvilinear, it is convenient to recall the established procedure and anticipate the consequences in two-dimensional space. The discussion will in fact be quite general for any sort of element in two or three dimensions, but it is very easy to draw in 2-D.

The first step is to write the integral form of the fundamental equation (3.12) as a summation of elements having volumes V and surfaces S. In two-dimensions of course these become areas and peripheral boundaries. Thus

$$\sum_{g=1,2,\ldots} \int_{V_g} \sigma^t \bar{\varepsilon}\, dV = \sum_{g=1,2,\ldots} \left(\int_{V_g} \mathbf{p}_v^t \bar{\mathbf{u}}\, dV + \int_{S_g} \mathbf{p}_s^t \bar{\mathbf{u}}\, ds \right). \qquad (5.1)$$

A FINITE ELEMENT PRIMER

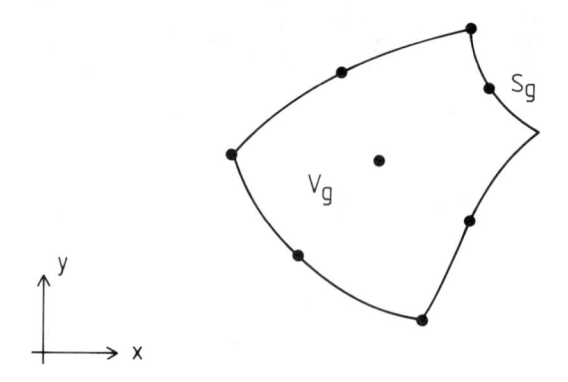

Figure 5.1

The next step is the original one. We have to find two-dimensional shape functions $N(x,y)$ which have unit values at one selected node, but zero at all other nodes on the element. The displacement unknowns at the nodes, $\mathbf{d}_g$ may be at the boundary of the element or inside. At this stage we spare the details and simply write

$$\begin{bmatrix} u \\ v \end{bmatrix} = \mathbf{u} = \mathbf{N}\mathbf{d}_g. \tag{5.2}$$

The rest should be routine. Enforce compatibility:

$$\varepsilon = \partial \mathbf{u} = \partial \mathbf{N}\mathbf{d}_g.$$

But $\mathbf{N}$ are assumed, known, shapes and therefore the differential operators can be applied to them. Putting these differentiated shapes as $\partial \mathbf{N} = \mathbf{B}$, we have

$$\varepsilon = \mathbf{B}\mathbf{d}_g. \tag{5.3}$$

Now the stress-strain law, ignoring thermal and other initial strains, is $\sigma = \mathbf{E}\varepsilon$. The above virtual work is therefore

$$\int_{V_g} \sigma^t \bar{\varepsilon}\, dV = \int_{V_g} \varepsilon^t \mathbf{E}\bar{\varepsilon}\, dV = \mathbf{d}_g^t \int_{V_g} \mathbf{B}^t \mathbf{E}\mathbf{B}\, dV \bar{\mathbf{d}}_g = \mathbf{d}_g^t \mathbf{k}_g \bar{\mathbf{d}}_g$$

where the element stiffness

$$\mathbf{k}_g = \int_{V_g} \mathbf{B}^t \mathbf{E}\mathbf{B}\, dV. \tag{5.4}$$

So far this looks just like the pinjointed framework route (recall equations (2.11) and (2.14)) except that now the material stiffness **E** is a 6×6 matrix, and **B** is an array of algebraic functions. The work done by the applied forces in (5.1) can be evaluated using $\mathbf{u} = \mathbf{N}\mathbf{d}_g$ and it becomes $\mathbf{P}_g^t \mathbf{d}_g$ where the kinematically equivalent forces emerge naturally as

$$\mathbf{P}_g = \int_{V_g} \mathbf{N}^t \mathbf{p}_v \, dV + \int_{S_g} \mathbf{N}^t \mathbf{p}_s \, ds. \tag{5.5}$$

For the last time it will be emphasised that these 'nodal forces' are simply *measures* of the magnitude of the distributed forces $\mathbf{p}_v$ and $\mathbf{p}_s$. They ensure that whatever the nature of $\mathbf{p}_v$ and $\mathbf{p}_s$ the correct work will be done over the selected displacement shape functions. There is no suggestion that physical concentrated forces are applied at the nodes.

Inserting (5.5) and (5.4) into (5.11) and summing using $\mathbf{d}_g = \mathbf{a}_g \mathbf{r}$

$$\left[\mathbf{r}^t \sum_g \mathbf{a}_g^t \mathbf{k} \mathbf{a}_g \right] \bar{\mathbf{r}} = \left[\sum_g \mathbf{P}_g^t \mathbf{a}_g \right] \bar{\mathbf{r}}$$

or **Kr = R**

where $\mathbf{K} = \sum \mathbf{a}_g^t \mathbf{k}_g \mathbf{a}_g$ and $\mathbf{R}^t = \sum_g \mathbf{P}_g^t \mathbf{a}_g$ again.

This process has been summarised once more to show the universality of the finite element method, but also to show now that we have been smarter than we think when we look into the details. We expect for example that the PVD finite element approximation will try to satisfy equilibrium as best it can, and this was proved in equation (3.14) which should, for an element, look like

$$\int_{V_g} \bar{\mathbf{u}}^t (\partial^t \boldsymbol{\sigma} + \mathbf{p}_v) \, dV - \int_{S_g} \bar{\mathbf{u}}^t [(\partial^t n) \boldsymbol{\sigma} - \mathbf{p}_s] \, ds = 0. \tag{5.6}$$

But this was proved using the Gauss Theorem which is only valid provided that all derivatives of $\boldsymbol{\sigma}$ and **u** are finite throughout V. We now have to concede that this may not be so. The vast majority of finite elements in use today do not achieve continuous stresses across interfaces, that is they violate equilibrium at a point. However, let us insist that *displacements* are continuous – such elements are said to be *conformal*. This means that the shape functions **N** are so chosen that when nodal displacements are matched at interfaces, then so will be the displacements

A FINITE ELEMENT PRIMER

$\mathbf{u} = \mathbf{N}\mathbf{d}_g$ across S. We can therefore say that equation (5.6) is true *within* a single element and up to its surface S_g, and consequently when the assembly summation (5.1) is carried out there will be contributions from adjacent elements ('1' and '2' in Figure 5.2) which have identical displacements $\mathbf{u}$ at their interface thus:

$$\int_{S_g} \bar{\mathbf{u}}^t [(\partial^t n)\sigma_1 - (\partial^t n)\sigma_2 - \mathbf{P}_s] \, ds = 0. \tag{5.7}$$

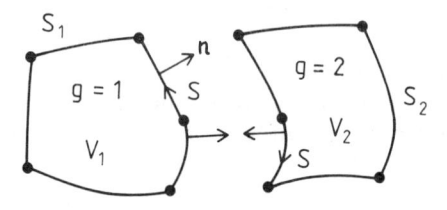

Figure 5.2

The different signs on $(\partial^t n)\sigma$ occur because 'n' is an outward positive normal and will consequently be in opposite directions for elements '1' and '2'. Equation (5.7) is another approximate satisfaction of equilibrium. If S_g is an internal interface and there are no applied loads ($\mathbf{p}_s = \mathbf{0}$) then (5.7) will attempt to ensure continuity of stresses *in the mean*, so that although the components of stress $(\partial^t n)\sigma$ may not balance at a point, the integrated residuals will vanish. Depending on the degrees of freedom in $\mathbf{u}$ there will be no out of balance force, or moment, or higher order equilibrium condition, on the whole element. In other words the displacement finite element method is capable of handling relaxed continuity in stresses between elements, in exactly the same way that it handled relaxed equilibrium conditions *within* the element.

Equation (5.7) is of course only true in that form if $\mathbf{u}$ is the same for continuous elements. There are several 'nonconforming' elements for which displacements are not contiguous save at the nodes themselves, and which perform very well. Indeed they seem to benefit from relaxed continuity on S_g in the form of sliding surfaces or rotational 'hinges'. A little flexibility is no bad thing in the 'overstiff' finite element solution. The behaviour of nonconforming elements is not so predictable since the absence of a 'weighted residual' form like (5.7) does not enable us to say that the residual gets smaller as the degrees of freedom in $\bar{\mathbf{u}}$ (the weight) increase as the mesh is refined. However there is a numerical 'Patch Test' which is a mathematically respectable alternative. A convenient shape, like a rectangle, has its interior divided into a nonsymmetrical (to avoid lucky answers) pattern of the element to be tested. One patch test then assumes a uniform strain over the rectangle, and applying the relevant boundary displacements checks that the finite element solution delivers a constant stress everywhere. (ref. 6).

5.2 Rectangular Elements

Rectangular elements are the easiest to discuss first and are very useful in their own right, particularly for stiffened thinwalled panel structures, and built up beams and boxes with little or no taper. Consider firstly the 4-noded element in Figure 5.3.

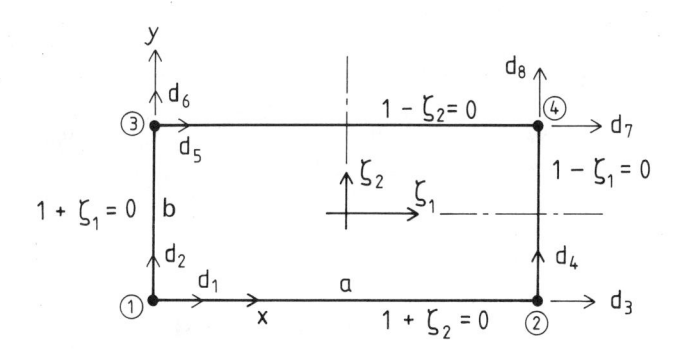

Figure 5.3

The required shape functions, having unit value at one node and zero values at others, can be written as

$$N_1 = (1-x/a)(1-y/b); \quad N_2 = (x/a)(1-y/b) \text{ etc.}$$

These *bilinear* shapes are obvious, and the idea of using the equations of the sides $(1-x/a=0$ etc.) to generate them will be extended. It is however much more convenient to transform to non-dimensional central coordinates $\zeta_1 = 2x/a - 1$ and $\zeta_2 = 2y/b - 1$ so that the element sides become $1 \pm \zeta_1 = 0$ and $1 \pm \zeta_2 = 0$. This change of variable has to be accommodated both in the strain derivatives ($\varepsilon = \partial \mathbf{u}$) and in the stiffness

A FINITE ELEMENT PRIMER

integral $k_g (dx = \frac{1}{2}ad\zeta_1)$. The required shape functions are then

$$N_1 = \tfrac{1}{4}(1-\zeta_1)(1-\zeta_2); \quad N_2 = \tfrac{1}{4}(1+\zeta_1)(1-\zeta_2);$$

$$N_3 = \tfrac{1}{4}(1-\zeta_1)(1+\zeta_2) \quad \text{and} \quad N_4 = \tfrac{1}{4}(1+\zeta_1)(1+\zeta_2)$$

whence

$$\mathbf{u} = \begin{bmatrix} u \\ v \end{bmatrix} = \begin{bmatrix} N_1 & 0 & N_2 & 0 & N_3 & 0 & N_4 & 0 \\ 0 & N_1 & 0 & N_2 & 0 & N_3 & 0 & N_4 \end{bmatrix}$$

where

$$\mathbf{d}_g^t = [d_1, d_2, d_3, \ldots, d_8].$$

The use of the above shape functions has one drawback – they are not *complete*. That is, all like powers in x and y are not present: in this case for example xy is present but x^2 and y^2 are not. Thus the variation of strain does not have the same order in all directions. A direct strain, say ε_{xx}, is constant in the 'x' direction whereas the shear varies linearly with x and y. Some improvement can be gained by diminishing the shear variation through the use of *reduced integration* (section 5.5) on the shear contribution in $\mathbf{k}_g$. Many users will prefer to turn to higher order elements rather than use a larger number of small 4-noded rectangles.

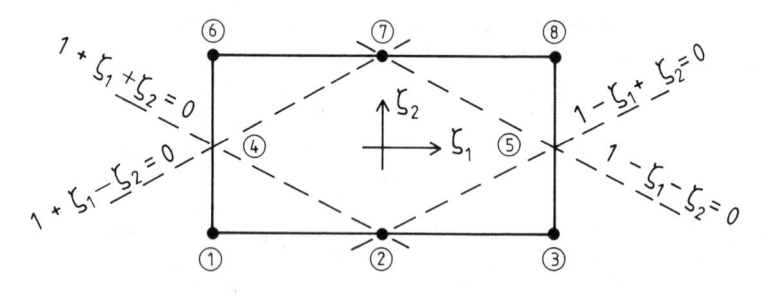

Figure 5.4

A very popular alternative is the higher order 8-noded element of Figure 5.4. (The equations of lines through new nodes is shown, the old ones are omitted for clarity.) Using the same trick of deploying the equations of these lines, the shape functions for the eight nodal values can be written as

$$N_1 = -\tfrac{1}{4}(1-\zeta_1)(1-\zeta_2)(1+\zeta_1+\zeta_2) \text{ etc.}$$

and

$$N_2 = \tfrac{1}{2}(1-\zeta_1)(1+\zeta_1)(1-\zeta_2) \text{ etc.}$$

(See (5.13) for details.)

The trouble taken to generate shapes which avoid producing zeros at the centre (0,0) is worth it. A 9-noded element including the centre produces a global stiffness matrix with inconveniently spread elements – see 'bandwidth' in section 8.

Having generated the shape functions **N**, the functions $\mathbf{B} = \partial \mathbf{N}$ follow and the element stiffness $\mathbf{k}_g$ (5.4) can be evaluated using the 2-D material stiffness:

$$\mathbf{E} = \frac{E}{1-v^2} \begin{bmatrix} 1 & v & 0 \\ v & 1 & 0 \\ 0 & 0 & \dfrac{1-v}{2} \end{bmatrix}.$$

However it is apparent that the product $\mathbf{B}^t\mathbf{E}\mathbf{B}$ in the stiffness integral is now becoming quite complicated even though the limits $-1 < \zeta_1, \zeta_2 < +1$ are simple. The situation becomes impossible when curvilinear sides are introduced and so in section 5.5 we show how to integrate the stiffness numerically. Even in the above rectangle, although the integration can be performed in closed form, most systems do not store the stiffness in this form in terms of a, b, **E**.

Both 4-node and 8-node rectangular elements are in popular use. The former has the simpler strain field and this is exploited in non-linear problems like elasto-plastic behaviour, where the stiffness has to be re-evaluated as the load increments are applied and plasticity spreads. But there is a danger in using too simple a description of stress and displacement fields and several of the NAFEMS benchmarks are examples where the dangers are revealed. One well known danger is the temptation to use slim rectangular elements as spars to represent slender beam bending.

The example in Figure 5.5 shows a uniform cantilever beam subjected to tip forces, the whole beam being divided into 3 elements. Only the maximum bending stress σ_{xx} on the upper surface is sketched for (a) a short cantilever and (b) a long one. Now high aspect ratio elements are to

A FINITE ELEMENT PRIMER

(a)

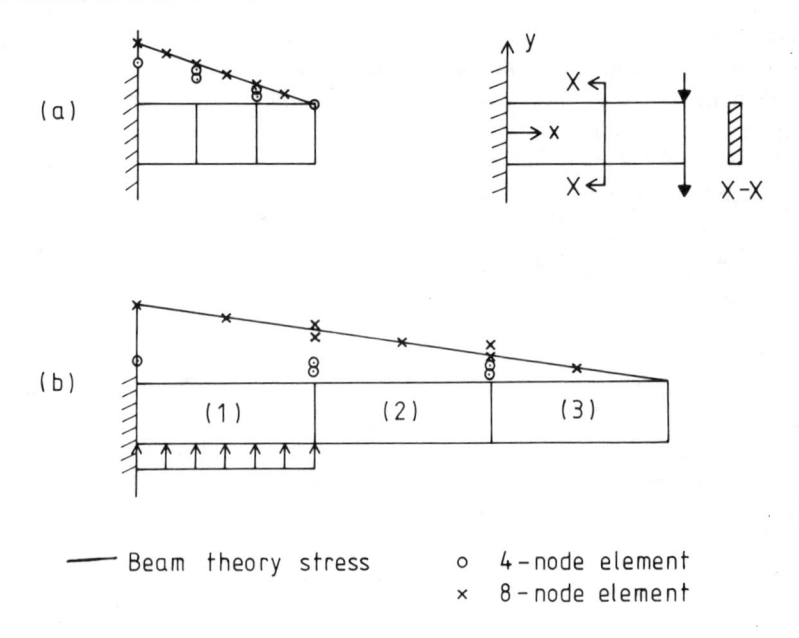

(b)

── Beam theory stress   o 4-node element
                                              × 8-node element

Figure 5.5

be discouraged: their stiffness matrix can become badly conditioned (det·**k** small) due to the fact that the displacement field is poorly defined by two nodal displacements which are close together. Nevertheless an aspect ratio of 3 is tolerable and this example should not be expected to misbehave on these grounds. Indeed the 8-node results in Figure 5.5(b) agree well with simple beam theory which itself should be accurate for case (b). The discontinuities which are a measure of accuracy are small, and if the arithmetic mean is taken, the nodal stresses are exact. Many commercial programs do output the arithmetic mean of nodal stresses at interfaces (irrespective of how many elements share the same node).

Turning to the 4-node results, they are a disaster, and if no other solution was available for comparison, they would be positively dangerous. Clearly not only is an explanation needed, but also some form of numerical diagnosis which will reveal that all is very far from well. The discontinuities in stress at interfaces are comparable with the mean values and should ring an alarm, even though they will not prepare the user for answers which are only 20% of the true values! However if suspicions are aroused, further investigation is always recommended, and another check is to see if stresses are in equilibrium with the applied loading, wherever this is

50

possible. These equilibrium checks confirm the poor answers. In Figure 5.5(b) the 'free' edge stress should be zero on the lower edge, but as shown on element (1) they are comparable with the solved stresses. These residual edge stresses cannot invalidate equilibrium for element (1) as a whole since this element has all rigid body freedoms and equilibrium will be satisfied. Therefore other edge stresses will be in error to compensate: for example the σ_{xx} stresses along the clamped support are significantly low and will not sum to a moment capable of balancing the cantilever tip-load unaided – another clue for the user.

The poor performance of this 4-noded element occurs because we have expected it to behave like a slender beam where bending stresses dominate over all other modes. The sketch in Figure 5.6(a) shows a pure bending mode (circular displacement curves and no shear deformation) whilst 5.6(b) shows the best that the 4-node linear-displacement model can manage.

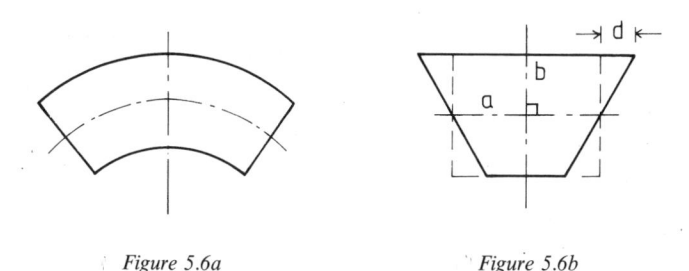

Figure 5.6a   *Figure 5.6b*

Only the centre of the 4-node element correctly has no shear deformation. At the element ends the shear strain ε_{xy}, is obviously d/b whereas the value of the 'bending stress' on the upper and lower surface is d/a. Thus the ratio of spurious shears to actual bending stresses is $\varepsilon_{xy}/\varepsilon_{xx} = a/b$, the aspect ratio. This explains the poor answers in Figure 5.5(a) and the disaster in 5.5(b). It is quite impossible for this element to simulate pure bending. When errors are involved we have seen that the finite element method will underestimate the strain energy and stresses will tend to be low – the shortfall in this pathological example is spectacular. This is one of the pitfalls of the displacement finite element method.

Several attempts have been made to introduce *equilibrium* finite elements, based on the principle of virtual forces, into commercial systems. These overestimate the strain energy and therefore give an upper bound which is generally on the safe side. One system uses both displacement and equilibrium methods and provides the user with both bounds to judge. However, pure equilibrium elements are difficult to construct for any but the simplest models, and further development in this field is unlikely to be commercial.

A compromise which should at least be mentioned in this Primer is the use of *hybrid* elements which are neither pure displacement models (compatibility exact, equilibrium approximate) nor pure equilibrium models (equilibrium exact, compatibility approximate). It is possible to mix the two approaches both inside an element and across interfaces. Of all the possible combinations the most significant is the use of element stress fields which satisfy equilibrium exactly in the interior but equilibrium is satisfied approximately 'in the mean' across interfaces – as in the displacement method. The principle of virtual forces generates an element flexibility which is then inverted to a stiffness, and doctored so that it looks like a conventional element stiffness to the system. There are now quite a few hybrid elements in finite element systems, and they tend to perform well since they lie somewhere between the lower and upper bounds mentioned previously. They also find stresses directly, instead of via strain which is a differentiated displacement in the displacement method, and not as accurate.

5.3 Triangular Elements

The triangular element was one of the very first approximate finite elements developed in 1956 and, before the advent of arbitrarily shaped isoparametric elements discussed in the next section, was one of the most widely used. It is a much more adaptable shape than the rectangle and it allows the user to tailor the element mesh in almost an infinite number of ways to suit any structural geometry or to anticipate any likely distribution of stress concentration. A large number of small elements can be crowded into a region of expected high stress gradients, and uniformly stressed regions can be left with a small number of larger triangles. This flexibility is not available with rectangular elements. (All this discussion of course presupposes that the engineer is so skilful at anticipating the answers that highly stressed regions can be densely packed. Or alternatively that adaptive mesh refinement can be afforded.)

The very first 3-noded triangle, with constant strain field, is still available in systems today. However it can be as poor a performer as the 4-node rectangle, and Figure 5.7 shows several problems.

TWO DIMENSIONAL MEMBRANES

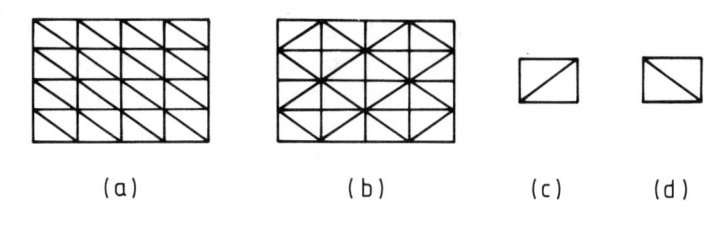

(a)　(b)　(c)　(d)

Figure 5.7

An assembly like 5.7(a) is clearly directionally sensitive, and this could be corrected by using the 'union jack' motif of 5.7(b) – which produces a larger bandwidth. This directional feature gets worse if we imagine the bending problem which so tested the simple rectangle. If the configuration 5.7(c) is placed in a pure bending field the upper triangle may be largely in horizontal tension with the lower one in compression. The consequent Poisson contractions in the vertical direction will then oppose each other instead of being the same. Some systems therefore build up a rectangle by adding (c) to (d) and forming a composite stiffness. The average strain components are then tolerable but the stresses need careful interpretation. The NAFEMS benchmark tests using triangular elements confirm that constant strain elements in a fine mesh are much inferior to higher order elements in a coarse one. It generally pays therefore to go to a higher order triangle with six nodes, but now to preserve symmetry it is easier to use a completely symmetrical coordinate system known as *area coordinates* shown in Figure 5.8.

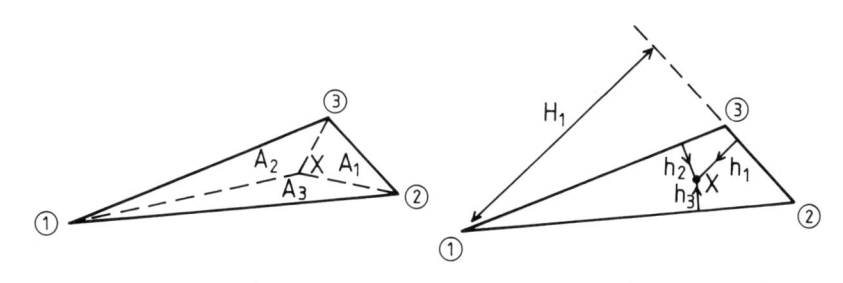

Figure 5.8

The position of any point X in the triangle is identified by the perpendicular distances h_1, h_2, h_3 from the three sides, and non-dimensionalised as

$$\zeta_1 = \frac{h_1}{H_1}; \; \zeta_2 = \frac{h_2}{H_2}; \; \zeta_3 = \frac{h_3}{H_3}.$$

53

A FINITE ELEMENT PRIMER

The term 'area coordinates' is used because $\zeta_1 = A_1/A$ etc. As $A = A_1 + A_2 + A_3$, $\zeta_1 + \zeta_2 + \zeta_3 = 1$ so the three coordinates are not independent. They do however satisfy our requirements that at node (1) $\zeta_1 = 1$ and $\zeta_2 = \zeta_3 = 0$ and so on. It is now straightforward to generate shape functions for any symmetrical distribution of nodes. For example the 6-noded triangle in Figure 5.9 has shown the equations of the three sides and the lines through the mid-side nodes.

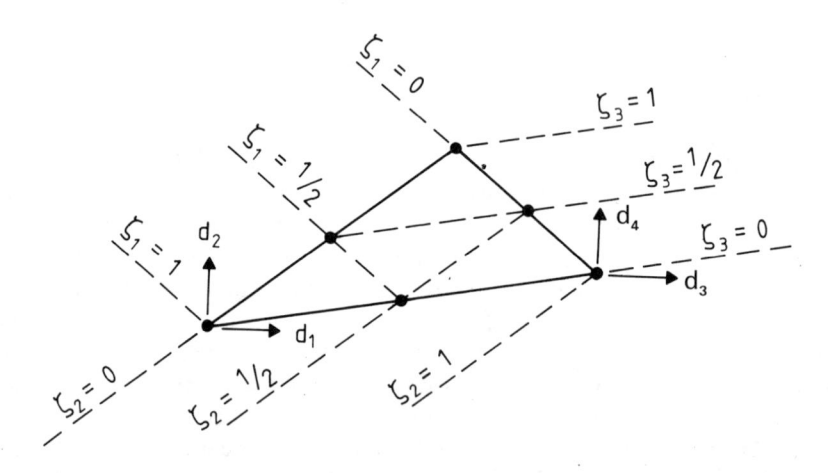

Figure 5.9

The shape functions are readily seen to be

$$N_1 = 2(\zeta_1 - \tfrac{1}{2})\zeta_1; \quad N_2 = 2(\zeta_2 - \tfrac{1}{2})\zeta_2; \quad N_3 = 2(\zeta_3 - \tfrac{1}{2})\zeta_3$$
$$N_4 = 4\zeta_1\zeta_2; \quad N_5 = 4\zeta_2\zeta_3; \quad N_6 = 4\zeta_1\zeta_3.$$

The displacements $\mathbf{u}^t = [u, v]$ in terms of $\mathbf{d}_g^t = [d_1, d_2, \ldots, d_{12}]$ are

$$\mathbf{u} = \begin{bmatrix} u \\ v \end{bmatrix} = \mathbf{N}\mathbf{d}_g$$

$$= \begin{bmatrix} N_1 & 0 & N_4 & 0 & N_2 & 0 & N_5 & 0 & N_3 & 0 & N_6 & 0 \\ 0 & N_1 & 0 & N_4 & 0 & N_2 & 0 & N_5 & 0 & N_3 & 0 & N_6 \end{bmatrix} \mathbf{d}_g.$$

(5.9)

To form $\mathbf{B} = \partial \mathbf{N}$ we need to find $\partial N/\partial x$ etc. so we must transform from x, y to ζ_i ($i = 1, 2, 3$). Now x and y are linear functions of ζ_i; and ζ_i of course have unit values at the three nodes $i=1,2,3$, so we must be able to write

$$x = \zeta_1 x_1 + \zeta_2 x_2 + \zeta_3 x_3$$
$$y = \zeta_1 y_1 + \zeta_2 y_2 + \zeta_3 y_3 \qquad (5.10)$$

where (x_i, y_i) refer to the coordinates of node $i=1,2,3$. Recalling that $\zeta_1 + \zeta_2 + \zeta_3 = 1$, any point in the triangle becomes

$$x = (x_1 - x_3)\zeta_1 + (x_2 - x_3)\zeta_2 + x_3$$
$$y = (y_1 - y_3)\zeta_1 + (y_2 - y_3)\zeta_2 + y_3.$$

It is now possible to form

$$\begin{bmatrix} \dfrac{\partial}{\partial x} \\ \dfrac{\partial}{\partial y} \end{bmatrix} = \frac{1}{A} \begin{bmatrix} y_2 - y_3 & y_3 - y_1 \\ x_3 - x_2 & x_1 - x_3 \end{bmatrix} \begin{bmatrix} \dfrac{\partial}{\partial \zeta_1} \\ \dfrac{\partial}{\partial \zeta_2} \end{bmatrix} \qquad (5.11)$$

where $A = (y_2 - y_3)(x_1 - x_3) - (y_1 - y_3)(x_2 - x_3)$ is twice the area of the triangle. The differentiation $\mathbf{B} = \partial \mathbf{N}$ can be performed using (5.11) on (5.9) and the element stiffness is again

$$\mathbf{k}_g = \int_A \mathbf{B}^t \mathbf{E} \mathbf{B} t \, dA \qquad (5.12)$$

where $dA = (dh_2)(dh_1 l_1 / H_1) = H_2 \, d\zeta_2 l_1 \, d\zeta_1 = A \, d\zeta_1 \, d\zeta_2$. The integrand in (5.12) is a formidable collection of products of derivatives of N with respect to ζ so that integration is frequently performed numerically. Some systems do however store the stiffness explicitly in terms of a, b, and $\mathbf{E}$ since, as it happens, the products can be integrated exactly.

$$\int_A \zeta_i^a \zeta_j^b \, dA = 2A \frac{a! b!}{(2 + a + b)!}.$$

The 6-noded triangle is quite popular and has no vices unless, like others, its aspect ratio exceeds 3 or 4. It has the virtue that the displacement field is *complete*, containing all possible products, so the element is truly isotropic; moreover it does this without having recourse to computationally inefficient internal nodes.

A FINITE ELEMENT PRIMER

One of the reasons why the triangle is not considered the universal panacea for awkward shapes has undoubtedly been the development of the *isoparametric element* introduced by Taig and Irons in the early sixties, and which enables us to use quadrilateral elements (and triangles) with curved sides. These shapes are extraordinarily versatile and have had a profound impact on the generation of meshes inside arbitrarily shaped regions. The concept of generating a curvilinear element has been so successful that the idea has also been extended to mesh generation of whole regions where a subregion is treated as a super-element to be subdivided later in a systematic way. (See section 7.)

5.4 The Isoparametric Curved Quad

The basic idea has already been used in a simpler way for the triangular element. If we examine (5.10) we were able to describe exactly the position of any point in the element in terms of the triangle's corner coordinates (x_i, y_i). The isoparametric 'mapping' simply carries this idea further and describes the geometry approximately and *in exactly the same way* as the displacement field, using the same shape functions. This democratic approach turns out to have many benefits, as we shall find. (The other options, where displacements and geometry are approximated using different shape functions have not been found to have much advantage.)

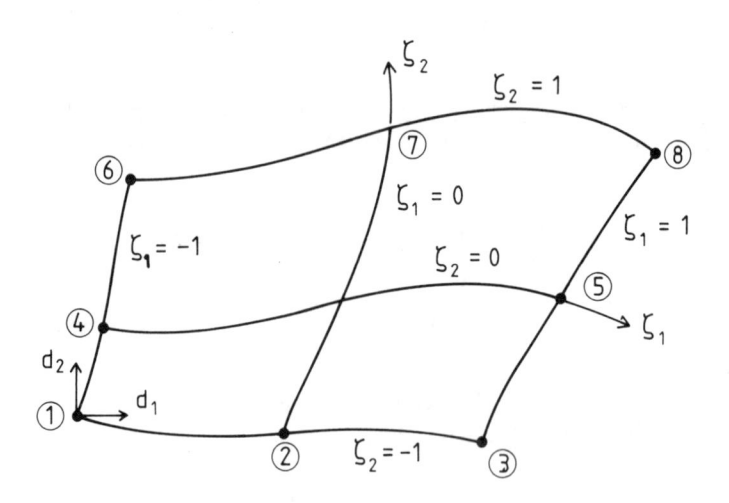

Figure 5.10

We concentrate on an 8-noded curvilinear quadratic element to illustrate the isoparametric approach. It is a popular element and we have already implied the shape functions in Figure 5.4. Using these again we simply express the x coordinates of any point inside the quad of Figure 5.10 as

$$x = N_1 x_1 + N_2 x_2 + N_3 x_3 + \cdots + N_8 x_8$$

where

$$N_1 = -\tfrac{1}{4}(1-\zeta_1)(1-\zeta_2)(1+\zeta_1+\zeta_2), \; N_2 = \tfrac{1}{2}(1-\zeta_1)(1+\zeta_1)(1-\zeta_2)$$

$$N_3 = -\tfrac{1}{4}(1+\zeta_1)(1-\zeta_2)(1-\zeta_1+\zeta_2), \; N_4 = \tfrac{1}{2}(1-\zeta_1)(1-\zeta_2)(1+\zeta_2)$$

$$N_5 = \tfrac{1}{2}(1+\zeta_1)(1-\zeta_2)(1+\zeta_2), \; N_6 = -\tfrac{1}{4}(1-\zeta_1)(1+\zeta_2)(1+\zeta_1-\zeta_2)$$

$$N_7 = \tfrac{1}{2}(1-\zeta_1)(1+\zeta_1)(1+\zeta_2), \; N_8 = -\tfrac{1}{4}(1+\zeta_1)(1+\zeta_2)(1-\zeta_1-\zeta_2) \quad (5.13)$$

and similarly for y.

Thus

$$\begin{bmatrix} x \\ y \end{bmatrix} = \mathbf{N}\mathbf{x} \quad (5.14)$$

where

$$\mathbf{x}^t = [x_1 \quad y_1 \quad x_2 \quad y_2 \quad \ldots \quad x_8 \quad y_8]$$

and

$$\mathbf{N} = \begin{bmatrix} N_1 & 0 & N_2 & 0 & \ldots & N_8 & 0 \\ 0 & N_1 & 0 & N_2 & \ldots & 0 & N_8 \end{bmatrix} \quad (5.15)$$

and of course

$$\begin{bmatrix} u \\ v \end{bmatrix} = \mathbf{u} = \mathbf{N}\mathbf{d}_g \quad (5.16)$$

where $\mathbf{d}^t = [d_1 \quad d_2 \quad d_3 \quad \ldots \quad d_{15} \quad d_{16}]$.

The term 'isoparametric' arises because (5.14) and (5.16) are of identical form, although to be pedantic the *shapes* are the same and not the parameters $\mathbf{x}$ and $\mathbf{u}$. We now require $\mathbf{B} = \partial \mathbf{N}$, but the operators ∂ are with respect to x and y and these must be transformed to match $\mathbf{N}(\zeta)$

according to the usual rules

$$\begin{bmatrix} \dfrac{\partial}{\partial \zeta_1} \\ \dfrac{\partial}{\partial \zeta_2} \end{bmatrix} = \mathbf{J} \begin{bmatrix} \dfrac{\partial}{\partial x} \\ \dfrac{\partial}{\partial y} \end{bmatrix} \tag{5.17}$$

where the two dimensional *Jacobian*

$$\mathbf{J} = \begin{bmatrix} \dfrac{\partial x}{\partial \zeta_1} & \dfrac{\partial y}{\partial \zeta_1} \\ \dfrac{\partial x}{\partial \zeta_2} & \dfrac{\partial y}{\partial \zeta_2} \end{bmatrix}. \tag{5.18}$$

The derivatives in (5.18) are found immediately from (5.14). Provided these are all finite we can invert (5.18) to get the desired operators:

$$\begin{bmatrix} \dfrac{\partial}{\partial x} \\ \dfrac{\partial}{\partial y} \end{bmatrix} = \mathbf{J}^{-1} \begin{bmatrix} \dfrac{\partial}{\partial \zeta_1} \\ \dfrac{\partial}{\partial \zeta_2} \end{bmatrix}. \tag{5.19}$$

The Jacobian is only a 2×2 matrix (3×3 in 3-D solids) and inversions can be performed immediately using (1.11) in this case. The adjoint is trivial, and $|\mathbf{J}|$ is needed anyway since the incremental area is given by $dA = |\mathbf{J}| d\zeta_1 d\zeta_2$ and

$$\mathbf{k}_g = \int_{-1}^{1} \int_{-1}^{1} \mathbf{B}^t \mathbf{E} \mathbf{B} |\mathbf{J}| t \, d\zeta_1 \, d\zeta_2. \tag{5.20}$$

So far we have demonstrated that mapping of displacement and geometry does provide a way to describe curvilinear shaped elements for which we can evaluate an element stiffness matrix. But several things, both good and bad, have to be stated.

It can be shown that in this isoparametric formulation the constant and linear terms in ζ became constant and linear displacement shapes in the real world of x and y. This is essential since we earlier mentioned that rigid body modes should be part of the assumed displacement field to guarantee element equilibrium, and that constant strain states are a necessity to ensure convergence as the element size becomes very small.

TWO DIMENSIONAL MEMBRANES

The power to create curvilinear shapes should not be abused, and curvilinear quads with extreme distortions from a rectangle must be avoided. It is quite possible in extreme cases to have two different points in the $(\zeta_1\text{-}\zeta_2)$ plane map to the same (x, y) or to have regions generated outside the element as in Figure 5.11.

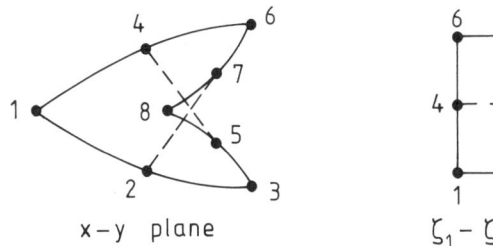

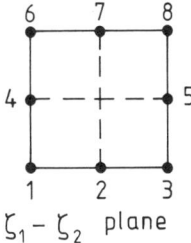

Figure 5.11

It can be shown that if **J** does not change sign inside the base square $-1 \leq \zeta_1, \zeta_2 \leq +1$ then the overlapping of Figure 5.11 is avoided. In a straight sided quad this simply means avoiding re-entrant angles. The determinant of **J** must not only not change sign, it must not be zero. This fact is obvious from (5.19) alone since even though $\mathbf{N}(\zeta)$ are differentiable with respect to ζ, the operators in ∂ will increase without limit as $|\mathbf{J}| \to 0$, and consequently produce infinite strains.

Unacceptable distortions can occur even in an undistorted element, if the other nodes are badly placed. Consider for example mapping a square $(a \times a)$ with the bottom edge along $0 < x < a$, node (1) at $x = 0$, node (3) at $x = a$, and for cussedness node (2) placed at $x = \beta a$ along the side. Using (5.13) along $\zeta_2 = -1$ we find the straight edge coordinates are given by

$$y = 0; \quad x/a = \beta - \beta\zeta_1^2 + \tfrac{1}{2}\zeta_1 + \tfrac{1}{2}\zeta_1^2.$$

This parametric equation looks innocent enough until we look at $\partial x/\partial \zeta_1$ which is plotted in Figure 5.12.

The gradient $\partial x/\partial \zeta_1$ is constant for $\beta = \tfrac{1}{2}$, that is when the internal node is at the midside. But if we move the node (2) in the direction of the corner (1) the gradient becomes variable and falls to zero at node (1) when $\beta = \tfrac{1}{4}$. Since $\partial y/\partial \zeta_1$ is zero along the entire side, the determinant of **J** (5.18) vanishes at node (1) and the stresses will be infinite. As a matter of

59

A FINITE ELEMENT PRIMER

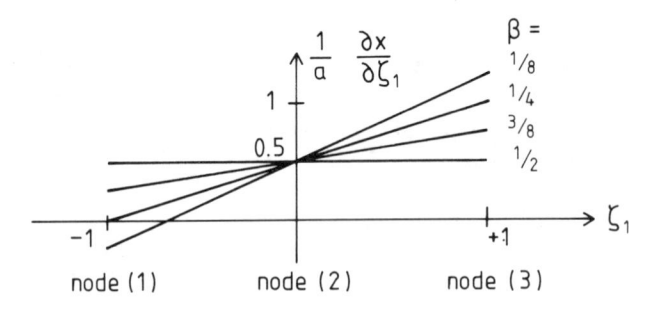

Figure 5.12

interest this feature is actually exploited in some systems to model the stresses at the tip of a crack $0 < x < a$, as a very cheap 'crack element' for fracture mechanics studies. (ref. 5).

As a guard against the above distortion some finite element systems will warn the user if a 'midside' node is placed outside the middle third of a side. Another safeguard is to recognise that all the above troubles stem from excessive variations in **J** and several systems will flag the user if this happens. This is a much more general diagnosis of distortion and the bounds placed on **J** seem to be based on experience.

Finally, typical results are shown in Figure 5.13 for a NAFEMS benchmark designed to test the accuracy of curved quads using respectable elements in an elliptical region which has to be generated by the system. A stress concentration occurs at the ends of the elliptical hole. One quarter of the doubly symmetrical structure is taken and divided into 6 and then 24 elements, and 4 and 8-node quads are used. The results for the stress along the inner boundary show fairly rapid convergence as the element size is halved, and the results even for 4-noded elements are reasonable provided the average of nodal stresses is taken. However, this smoothing clearly does not work at the position of maximum stress; the only way to find peak stresses accurately is to use a fine enough mesh or a high order element. It is usually claimed that fewer higher order elements is the better option: this is certainly true in this example.

Stress discontinuities are not a complete guide to accuracy in the problem, where most of the error in the maximum stress for the superior 8-node elements arises in the approximate way that the preprocessor generates the curvature of the inner boundary.

elliptic 'annulus' edge stresses

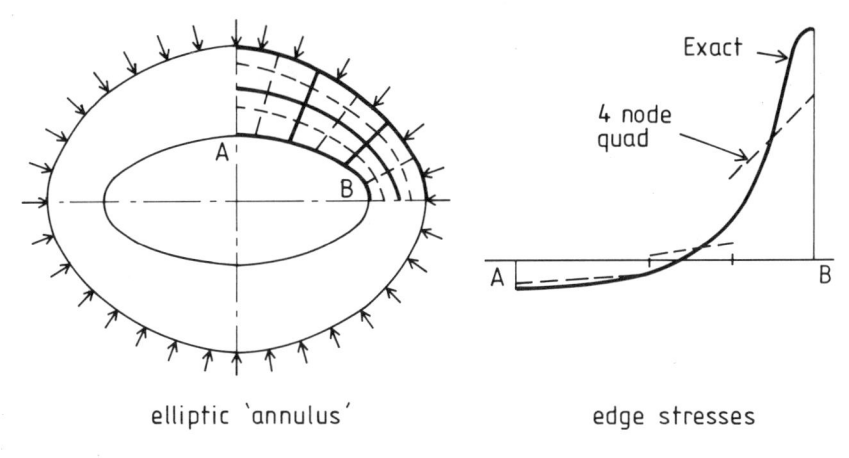

Figure 5.13

5.5 Numerical Integration

We have seen how the element stiffness integrals become progressively more complicated. The differentiated shape functions themselves, $\mathbf{B} = \partial \mathbf{N}$, grew algebraically more cumbersome as the order of the displacement field increased. In the case of isoparametric curved elements the mapping from (x, y) to (ζ_1, ζ_2) forced us to introduce $|\mathbf{J}|$ into equation (5.20) which finally destroyed any chances of evaluating the integrals in closed form. We now discuss very briefly the most commonly used form of evaluating the stiffness integrals numerically – *Gaussian Quadrature*, or simply Gaussian integration. We can safely assume that the element has been mapped into a ζ-plane with limits $-1 < \zeta < +1$ or, if the element shape is simple like a beam or a rectangle, then a change of variable to the same limits is trivial. Consider first then a one dimensional problem where the integrand $\mathbf{B}^t\mathbf{EB}$ is called, for convenience, a function $f(\zeta)$.

The simplest form of numerical integration, familiar from schooldays, finds the area under the curve $f(\zeta)$ by dividing the axis $-1 < \zeta < +1$ into equal segments, multiplies the local value of $f(\zeta)$ in the middle of each segment by the length of the segment, and sums all these small rectangular areas. This is the crudest approximation to the area under a curve. A much better way relies on the concept that any shape $f(\zeta)$ can be represented approximately over the interval $-1 < \zeta < +1$ by a polynomial (of order 'n' say) which can then of course be integrated exactly. A polynomial of order n can be 'fitted' (have its coefficients chosen) to $f(\zeta)$ by imposing the values of $f(\zeta_i)$ at $n+1$ points $\zeta_0, \zeta_1, \zeta_2, \ldots, \zeta_n$. A parabola ($n=2$) can be made to pass through 3 points for example. Simpson's rule and the more general Newton–Cotes formulae use this approach (see ref. 1). These formulae all produce a summation as the approximation for the definite integral

$$\int_{-1}^{1} f(\zeta) \, d\zeta = \sum_{i=0}^{n} w_i f_i.$$

Here f_i are all the $n+1$ function values $f(\zeta_1)$ from $f_0 = f(-1)$ up to $f_n = f(+1)$, and w_i are *weights* which depend on the order of the polynomial fit (for Simpson's rule $n=2$, $w_0 = w_2 = 1/3$, $w_1 = 4/3$). The Gauss method does not use equal intervals but instead chooses both the weights w_i *and* the sampling points $\zeta_0, \zeta_1, \ldots, \zeta_n$ as variables in such a way that the summation is exact up to the chosen order. With twice the number of disposable variables ζ_i and w_i, a polynomial of order $2n+1$ can now be integrated exactly. The particular positions of these sampling points are known as *Gauss Points*. (They turn out to be the roots of Legendre Polynominals and so the method is occasionally referred to as a Gauss–Legendre weighting scheme.) These Gauss Points are symmetrically placed about $\zeta = 0$ and, together with the weights w_i can be shown to have the values in the accompanying table.

Table 5.1 Gauss Points and Weights

Number of Gauss Points	Order of polynomial integrated exactly	Gauss Points	Weights
1	1	0	2.0
2	3	± 0.57735 ($1/\sqrt{3}$)	1.0
3	5	0	8/9
		± 0.77459 ($\sqrt{3/5}$)	5/9
4	7	± 0.33998	0.65215
		± 0.86114	0.34785

This form of integration is particularly efficient for stiffness integrals because it involves only a half of the sample values f_i needed in a conventional Newton–Cotes type summation. The only penalty paid is the awkward positions of the Gauss Points, but the function B^tEB is likely to be complex in form and it is far easier to evaluate it at irregularly spaced points than to do it at twice the number of regularly spaced points to achieve comparable accuracy. Moreover there is an added bonus in having to find these Gauss Points since it was shown by Barlow (ref. 8) in the sixties that these points are optimum sampling points for examining stresses, in some elements. Intuitive reasoning suggests that these points should be good representations since they have been chosen as local mean values in an integral, and the finite element method has been shown to satisfy approximately 'in the mean'. For example in the beam element in Figure 4.3, the element stiffness integral contains $(\mathbf{N}'')^t\mathbf{N}''$ which is quadratic and requires two Gauss points for exact evaluation. Turning to the example of Figure 4.5a, these two Gauss points in every element are indeed the intersection of the linear finite element solutions with the exact parabolic bending moment. They are optimal sampling points in one dimension.

The one dimensional version of Gaussian Quadrature is readily extended to two dimensions if we use the rectangles or isoparametric quads of sections 5.2 and 5.4. For example the 8-node quad employs displacement fields quadratic in ζ_1 and ζ_2 as are the stresses. The product $\mathbf{B}^t\mathbf{EB}$ is therefore fourth order and it would seem necessary to use three Gauss points in both ζ_1 and ζ_2 directions – nine points in all. (It is possible to look at the total order 'm' in products $\zeta_1^n \zeta_2^{m-n}$ and integrate with fewer points distributed in a different way to 3×3, but this coding is rarely done). A very popular dodge is to use *reduced integration* at fewer points than necessary with the aim of decreasing the stiffness and so compensating for the overstiff finite element model – it is also cheaper of course. Some doubts have been expressed at the wisdom of emulating a slightly lower order element whilst still retaining the full degrees of freedom and undoubtedly a singular stiffness matrix will result using a 2×2 scheme for a single 8-node element even if it is supported conventionally. But in practice, larger collections of elements will not suffer from such mechanisms, and 2×2 integration for this popular element has nowadays become the norm. The optimal sampling points for this element turn out to be these same 2×2 points and not the higher order 3×3, so the advantages of reduced integration are compounded. Some systems which use Gauss point stresses will extrapolate to the element nodes and then output the mean value if several elements meet at a node; but a little care is needed. Maximum stresses usually occur at edges of plates or at other discontinuities which will be element boundaries. It is possible to miss these peak stresses if the elements in this region are too large. However for constructing internal stress contours, Gauss point values are ideal.

A FINITE ELEMENT PRIMER

The beam problem of Figure 5.5 conveniently illustrates Gauss point virtues. The 8-node elements predicted the linearly varying bending stresses σ_{xx} very well. The shear σ_{xy} and transverse stresses σ_{yy} are not so well predicted due to the fact that in this particular problem σ_{yy} should vary antisymmetrically with y and be zero on both upper and lower edges. Unfortunately this element has only a linear variation of σ_{yy} with y (a cubic is required) and so the PVD tries to suppress it and leaves both transverse components σ_{yy} and shears σ_{xy} almost constant through the depth. A higher order element, or two elements through the depth, would cure this problem. The variation of shear along the beam is therefore a poor approximation as Figure 5.14 shows (see ref. 1 also; p. 283) but the Gauss point stresses predict the *mean* values of shear very accurately confirming that these points are an optimum measure of mean values. The shear stress actually varies parabolically through the depth and has a maximum value 50% greater than the mean, but to capture this would need the extra element or degrees of freedom.

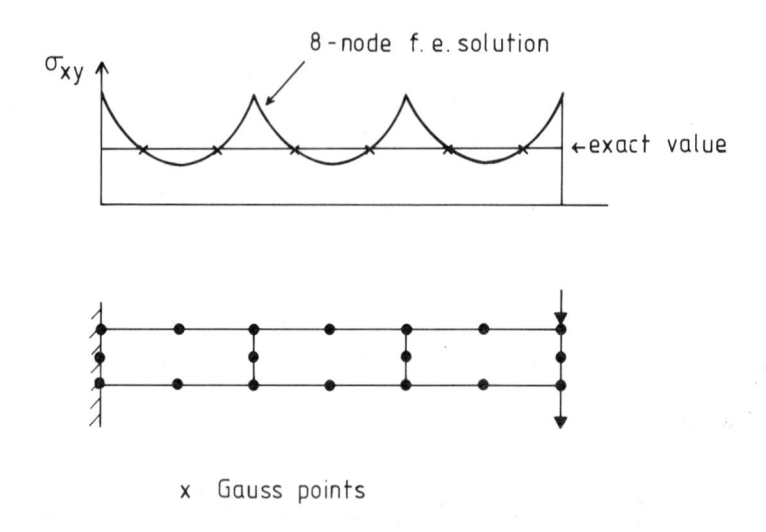

x Gauss points

Shear stresses for problem of fig. 5.5

Figure 5.14

In practice the user will not know with confidence how inadequate the model is. The finite element system should be relied upon to warn users that answers are suspect, and this example illustrates one of the difficulties.

It is proper to look at equilibrium errors across element interfaces or on boundaries such as here. The residual errors could then be compared with the maximum stress in the local element to ascertain whether they are serious. But here the bending stresses σ_{xx} are solved very accurately and are much greater than the edge errors in σ_{yy} or σ_{xy}, so the solution would normally be acceptable. On the other hand these errors could be serious for structures which may be potentially weak in shear σ_{xy}, or transverse σ_{yy}. (Composite beams of plastic or concrete for example can be weak in shear in the transverse direction.) So the failure criterion should be incorporated into the automated checks if possible.

5.6 Initial Strains

Neither numerical integration nor thermal strains are peculiar to two dimensional membrane structures, but it is convenient to illustrate using 2-D so they are included in this section. Initial strains are those due to causes other than stress, and some confusion exists here since thermal strains or 'lack of fit' may themselves cause stresses unless the structure is statically determinate (where stresses can depend only on the applied loads). The situation is really quite straightforward if we look at strains and not stresses first, recalling that the total strain, $\varepsilon = \partial \mathbf{u}$, is due to any cause whatsoever. The assumption made before equation (3.11) is repeated, namely that strain is proportional to stress, and that in the absence of stress the initial strains are η. Thus when both are present

$$\varepsilon = \mathbf{f}\sigma + \eta. \tag{5.21}$$

The inverse of this is

$$\sigma = \mathbf{E}(\varepsilon - \eta). \tag{5.22}$$

The routine is now exactly as before, so that the fundamental equation (3.12) becomes, for all elements:

$$\sum_g \int_{V_g} (\varepsilon - \eta)^t \mathbf{E}\bar{\varepsilon}\, dV = \sum_g \left(\int_{V_g} \mathbf{p}_v^t \bar{\mathbf{u}}\, dV + \int_{S_g} \mathbf{p}_s^t \bar{\mathbf{u}}\, ds \right).$$

Putting $\mathbf{u} = \mathbf{N}\mathbf{d}_g$ and $\varepsilon = \partial \mathbf{u} = \mathbf{B}\mathbf{d}_g$,

$$\sum_g \left(\mathbf{d}_g^t \int_{V_g} \mathbf{B}^t \mathbf{E}\mathbf{B}\, dV - \int_{V_g} \eta^t \mathbf{E}\mathbf{B}\, dV \right) \bar{\mathbf{d}}_g = \sum_g \left(\int_{V_g} \mathbf{p}_v^t \mathbf{N}\, dV + \int_{S_g} \mathbf{p}_s^t \mathbf{N}\, ds \right) \bar{\mathbf{d}}_g.$$

Comparing with equations (5.1) and (5.4) the only addition is obviously the term in $\boldsymbol{\eta}$, which can be transferred to the right-hand side and placed with the mechanical loading as coefficients of $\bar{\mathbf{d}}_g$. The kinematically equivalent forces (5.5) now become therefore

$$\mathbf{P}_g = \int_{V_g} \mathbf{N}^t \mathbf{p}_v \, dV + \int_{S_g} \mathbf{N}^t \mathbf{p}_s \, ds + \int_{V_g} \mathbf{B}^t \mathbf{E} \boldsymbol{\eta} \, dV. \tag{5.23}$$

Provided the third integral can be evaluated, the mechanical loading and thermal loading are grouped together, and the complete system, $\mathbf{Kr} = \mathbf{R}$, solved as usual. The recovery of element strains from $\mathbf{r}$ is also as usual ($\boldsymbol{\varepsilon} = \mathbf{Bd} = \mathbf{Ba}_g \mathbf{r}$) but remember that the stress now has to be found using (5.22).

One of the commonest forms of initial strain is due to temperature T, causing initial strains which are proportional to T over a limited range. Thus $\boldsymbol{\eta} = \boldsymbol{\alpha} T$ where $\boldsymbol{\alpha}$ is a column of thermal expansion coefficients. Most systems evaluate the kinematically equivalent thermal loading in (5.23) by expressing the temperature $T(s, y, z)$ in an element in terms of the temperatures at the nodes $\mathbf{T}_N$, using *the same* interpolations as the displacement. That is,

$$T(x, y, z) = \mathbf{NT}_N \quad \text{and} \quad \boldsymbol{\eta} = \boldsymbol{\alpha} \mathbf{NT}_N \tag{5.24}$$

whence

$$\mathbf{P}_g = \int_{V_g} \mathbf{B}^t \mathbf{E} \boldsymbol{\alpha} \mathbf{NT}_N \, dV. \tag{5.25}$$

This approach is not consistent since the approximation for the thermal strains (5.24) is the same as the displacement field whereas the strains, $\boldsymbol{\varepsilon} = \partial \mathbf{Nd}_g = \mathbf{Bd}_g$, are a lower order. Some finite element systems therefore approximate the temperature field by a lower order expansion such as fitting it through fewer nodes – or better by using a least squares fit and matching the temperature to the Gauss point values. In this case both types of strain, $\boldsymbol{\varepsilon}$ and $\boldsymbol{\eta}$, are of the same order and the stresses in (5.22) are consistent. Only the temperature field has been approximated in a cruder fashion – and no diagnostics will reveal the consequences of this. It seems that possibly the best compromise is to expand the temperature field using the same interpolations and all nodal points as the displacements, but to sample the stresses at a reduced number of Gauss points as shown on the following NAFEMS benchmark, Figure 5.15.

TWO DIMENSIONAL MEMBRANES

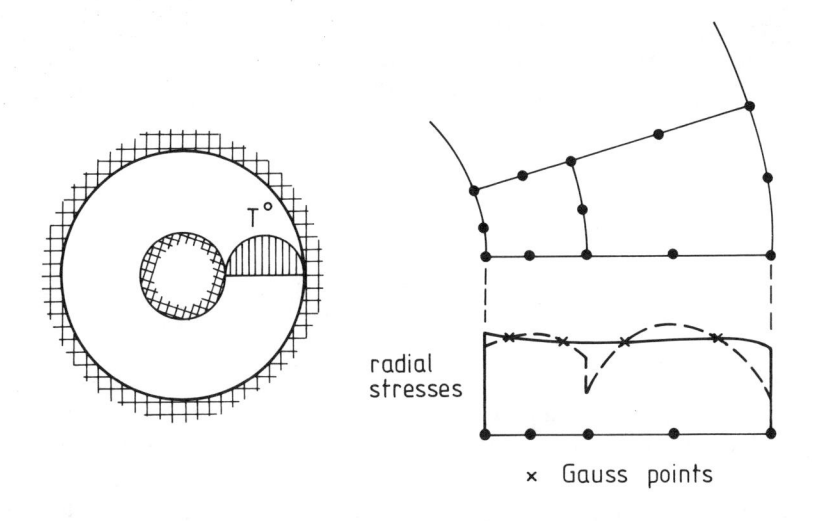

Figure 5.15   Clamped Annulus : Parabolic Temperature

6. Bricks, Plates and Shells

6.1 Introduction

Chapter 5 was a long one, with many diversions into isoparametric formulation, numerical and reduced integration, optimal stress sampling, thermal stresses and the like. None of these special topics is particular to a two-dimensional membrane which was the purpose of the chapter. However it was felt that it was easier and more convincing to discuss these features in the light of real element formulations, and the two-dimensional membrane formulation is the simplest one which contains all these finite element manoeuvres. This sixth chapter is devoted to a discussion of the other forms of element used today, in addition to the ubiquitous beams and membranes (which incidentally account for more than 80% of the structures analysed commercially). It will not be necessary to reformulate principles nor to delve deeply into algebraic detail since this would largely be a repetition of previous discussions extended to more complex elements. This decision is both expedient and fortunate since plate elements (flat, thin, and loaded normally to their surface) and shell elements (ditto, but curved) are not simple. There are both geometrical complexities and physical ones which are not present in membranes and which make plates and shells formidable elements to formulate and often expensive to use. The most expensive elements of all are the three-dimensional solids, and yet these are the easiest conceptually to discuss so we start with them.

6.2 Solid Elements

The 20-noded curved 'brick' element shown in Figure 6.1 is a logical extension of the popular 8-noded membrane element first shown in Figure 5.10. The shape functions are generated in exactly the same way using the equations of the sides in curvilinear coordinates ζ_1, ζ_2 and ζ_3. Thus the shape function for N_{19} for example is

$$N_{19} = \tfrac{1}{4}(1-\zeta_1)(1+\zeta_1)(1+\zeta_2)(1+\zeta_3)$$

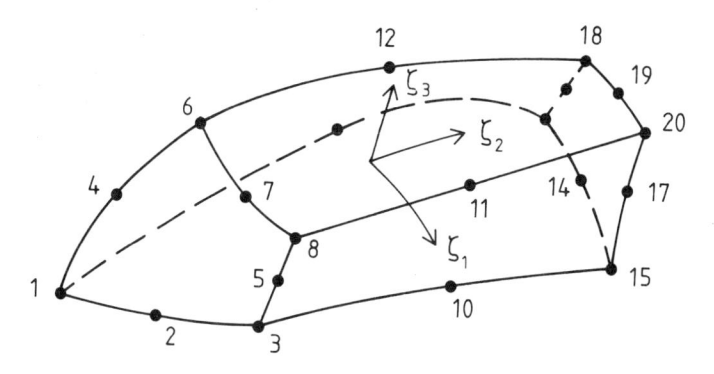

Figure 6.1 The curvilinear brick.

giving unity at the node 19 and zero at all other 19 nodes. The corner nodes are equally straightforward, so for example

$$N_{20} = \tfrac{1}{8}(1+\zeta_1)(1+\zeta_2)(1+\zeta_3)(\zeta_1+\zeta_2+\zeta_3-2).$$

The stiffness integral is the same as (5.20) except that the infinitesimal volume in the basic coordinates is $d\zeta_1, d\zeta_2, d\zeta_3$ and so

$$\mathbf{k}_g = \int_{-1}^{1} \int_{-1}^{1} \int_{-1}^{1} \mathbf{B}^t \mathbf{E} \mathbf{B} |\mathbf{J}| \, d\zeta_1, d\zeta_2, d\zeta_3 \qquad (6.1)$$

where $|\mathbf{J}|$ is now of course a 3×3 determinant. Full $3 \times 3 \times 3$ Gaussian integration is extremely expensive and reduced integration is absolutely necessary for these elements to be viable. However reduced integration has its pitfalls as we mentioned in chapter 5. Some – usually higher order – strains can be unfortunately zero at the Gauss Points in some modes and so the element simulates a lower order model but with the extra degrees of freedom, and hence $\mathbf{k}_g$ is singular. A crude, but not infallible way of looking at this problem is to recognise that the stiffness matrix integral (6.1) is receiving its information on the type of element solely from the Gauss points and this must reflect the true degrees of freedom of the element. Consider for example the 8-node rectangle deformed as shown in Figure 6.2, where it can be verified that at all four reduced-integration Gauss Points there are zero strain components.

In the case of this two-dimensional element with three strain components $(\varepsilon_{xx}, \varepsilon_{yy}, \varepsilon_{xy})$ the work integral has 4×3 contributions whereas the single element, supported to prevent three rigid body movements, has $8 \times 2 - 3$

A FINITE ELEMENT PRIMER

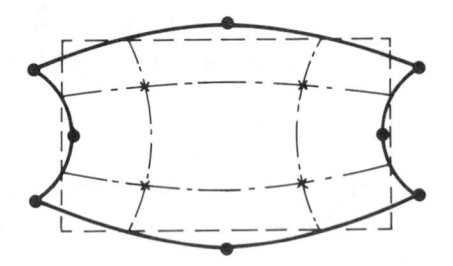

Figure 6.2 Zero Energy mode. *Gauss Points.

= 13 degrees of freedom. This exceeds 12 so a single kinematic mechanism is possible. However if two elements are joined together the total work integral has 24 contributions but there are only 13 nodes and 26 freedoms, less three rigid body nodes leaving 23, and so the structure is respectable. This incidentally illustrates one of the cases against relying entirely on single element tests – namely that when suspect elements are gathered together in friendly groups they form respectable structures. The same 'hourglass' mode can happen in a 20-node brick clearly, but it will also persist if a row of bricks is laid end-to-end to form a solid rod. Users obviously have to exercise greater care in three dimensions, and in the end may have to rely on the system's diagnostics which should recognise a poor conditioning number as described in chapter 8.

A 10-noded tetrahedron can be generated in a fashion similar to the 6-noded triangle, but tetrahedra are becoming less popular since it is costly to check that none are missing in complex three-dimensional shapes. In fact some systems will firstly form brick elements in an automated fashion from tetrahedra to guarantee no holes – but this rather defeats the original purpose. They are sometimes used to fill awkward shapes or as transitions from a coarse mesh to a fine. But again it is possible to do this using bricks and without unacceptable distortion as Figure 6.3 shows in two dimensions.

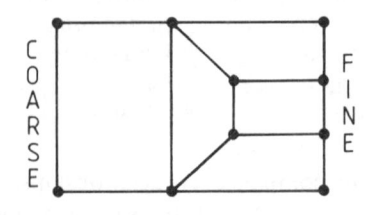

Figure 6.3 Transition from two nodes to four.

A 'wedge' element or prism, consisting of a triangle in two dimensions and quadrilateral in the third, is a more popular 15-noded element for filling in corners.

One thing these three-dimensional elements have in common is that they are expensive, and the user should resist the temptation to deploy them simply because they look straightforward and versatile. With modern interactive preprocessors and mesh generators it is very easy to generate a large number of nodes very quickly and a large budget deficit at the same time. If a two-dimensional approximation can be safely made then it will be significantly cheaper. This is the rationale behind plate bending and shell elements of course, but before turning to these we discuss the very common axisymmetrical structure where all sections through a central axis look identical. Pressure vessel technology often leads to axisymmetrical structures, and axisymmetrical loading as well. So do many other branches of engineering whether mechanical, civil, marine, aerospace and chemical: and they may be solid or thinwalled shells. Any two dimensional membrane element can be converted to an axisymmetrical model; we simply illustrate the quad in Figure 6.4 as one of the most widely used.

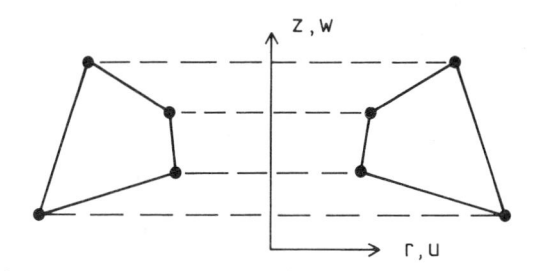

Figure 6.4 Axisymmetrical element.

The problem is genuinely two-dimensional with two displacement components u and w, but with four strain components

$$\boldsymbol{\varepsilon}^t = [\varepsilon_{rr}, \varepsilon_{\theta\theta}, \varepsilon_{zz}, \varepsilon_{zr}].$$

The strain-displacement compatibility relationships, $\boldsymbol{\varepsilon} = \partial \mathbf{u}$ (3.9) in the physical space variables are slightly different in cylindrical polar coordinates:

A FINITE ELEMENT PRIMER

$$\varepsilon = \begin{bmatrix} \frac{\partial}{\partial r} & 0 \\ \frac{1}{r} & 0 \\ 0 & \frac{\partial}{\partial z} \\ \frac{\partial}{\partial z} & \frac{\partial}{\partial r} \end{bmatrix} \begin{bmatrix} u \\ w \end{bmatrix} = \partial \mathbf{u}. \tag{6.2}$$

So if the geometry (r, z) and the displacement field (u, w) are expanded as usual ($\mathbf{u} = \mathbf{Nd}$) then $\varepsilon = \partial \mathbf{Nd} = \mathbf{Bd}$ produces a matrix $\mathbf{B}$ slightly different to the rectilinear elements. The elemental area in the stiffness (5.20) is also slightly different.

$$\mathbf{k}_g = \int_{-1}^{1} \int_{-1}^{1} \int_{-\pi}^{\pi} \mathbf{B}^t \mathbf{E} \mathbf{B} r \, d\theta |\mathbf{J}| d\zeta_1 \, d\zeta_2. \tag{6.3}$$

Some of the terms in $\mathbf{B}$ arising from (6.2) have $1/r$ as a multiplier and this may need special treatment on the central axis, $r = 0$. However if the isoparametric formulation is used with Gaussian quadrature then no integration point is on this axis. If further the stresses are sampled at the Gauss points and extrapolated to the central axis, all difficulties on $r = 0$ are avoided. Otherwise special central axis elements are needed with no radial displacements on the central axis, and the value of $\varepsilon_{\theta\theta}$ is equal to ε_{rr} on this axis anyway in the axisymmetric loading case.

Frequently axisymmetrical solids are subjected to non-symmetrical loading, and in a few systems the user has to resort again to general bricks. The geometry of course can still be generated taking advantage of symmetry in the first instance. However many systems can analyse axisymmetrical solids and shells, subject to general loads, by expanding the known loading and the unknown displacements as Fourier Series in the circumferential coordinate. It is well known that any periodic function (say the body forces $\mathbf{p}_v$) can be expanded as a Fourier series with symmetric and antisymmetric harmonics

$$\mathbf{p}_v = \sum_{n=0}^{\infty} \mathbf{p}_{vn} \cos n\theta + \sum_{n=1}^{\infty} \mathbf{p}'_{vn} \sin n\theta \tag{6.4}$$

where

$$\mathbf{p}_{v0} = \frac{1}{2\pi} \int_{-\pi}^{\pi} \mathbf{p}_v \, d\theta; \quad \mathbf{p}_{vn} = \frac{1}{\pi} \int_{-\pi}^{\pi} \mathbf{p}_v \cos n\theta \, d\theta; \quad \mathbf{p}'_{vn} = \frac{1}{\pi} \int_{-\pi}^{\pi} \mathbf{p}_v \sin n\theta.$$

Thus if cylindrical coordinates (r, θ, z) are used, the body forces can be written as

$$\mathbf{p}_v^t = [p_r \quad p_\theta \quad p_z]$$

where

$$p_r = \sum p_{rn} \cos n\theta + \sum p'_{rn} \sin n\theta,$$

$$p_\theta = \sum p_{\theta n} \sin n\theta + \sum p'_{\theta n} \cos n\theta,$$

and

$$p_{r0} = \frac{1}{2\pi} \int_{-\pi}^{\pi} p_r \, d\theta; \quad p_{rn} = \frac{1}{\pi} \int_{-\pi}^{\pi} p_r \cos n\theta \, d\theta; \quad p'_{rn} = \frac{1}{\pi} \int_{-\pi}^{\pi} p_r \sin n\theta \, d\theta, \quad (6.4)$$

and so on.

Many loadings can be approximated quite accurately with only a few harmonics, such as in Figure 6.5. For example large scale shell buildings, cooling towers, storage tanks and so on may be subjected to wind loadings. Other moving transport vehicles or components will be subjected to gravity or other inertia loadings which will not be axisymmetrical. In such problems it may be necessary to use only two or three harmonics, and even in problems where the forces are applied only to portions of the surface it is possible that $\mathbf{p}_{vn}$ and $\mathbf{p}'_{vn}$ may become insignificant for n greater than 10. If this is not so then it implies that the loading is not smooth, and concentrated forces or constraints are present. In this case a general three-dimensional analysis may be the only alternative.

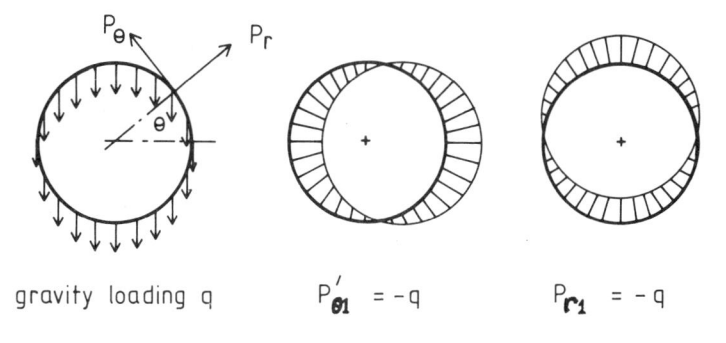

Figure 6.5 Single harmonic loading on cylinders.

A FINITE ELEMENT PRIMER

The unknown displacements are now expanded in a similar fashion using cylindrical coordinates and recognising that now all six stress components are present.

$$\begin{bmatrix} \varepsilon_{rr} \\ \varepsilon_{\theta\theta} \\ \varepsilon_{zz} \\ \varepsilon_{zr} \\ \varepsilon_{r\theta} \\ \varepsilon_{\theta z} \end{bmatrix} = \begin{bmatrix} \dfrac{\partial}{\partial r} & 0 & 0 \\ \dfrac{1}{r} & \dfrac{1}{r}\dfrac{\partial}{\partial \theta} & 0 \\ 0 & 0 & \dfrac{\partial}{\partial z} \\ \dfrac{\partial}{\partial z} & 0 & \dfrac{\partial}{\partial r} \\ \dfrac{1}{r}\dfrac{\partial}{\partial \theta} & \dfrac{\partial}{\partial r} - \dfrac{1}{r} & 0 \\ 0 & \dfrac{\partial}{\partial z} & \dfrac{1}{r}\dfrac{\partial}{\partial \theta} \end{bmatrix} \begin{bmatrix} u \\ v \\ w \end{bmatrix}. \quad (6.5)$$

The displacements are

$$u = \sum u_n \cos n\theta + \sum u'_n \sin n\theta$$
$$v = \sum v_n \sin n\theta + \sum v'_n \cos n\theta \quad (6.6)$$
$$w = \sum w_n \cos n\theta + \sum w'_n \sin n\theta.$$

It is found that if (6.6) is substituted into (6.5), and then the strains converted to stresses via the material stiffness **E**, then when these stresses are substituted into the PVD a separate equation emerges for each harmonic term $\cos n\theta$ or $\sin n\theta$. These simplifications occur because the work integral contains products like $\cos n\theta \cos m\theta$ which vanish when integrated from $\theta = -\pi$ to $+\pi$, unless $n = m$. (The terms in a Fourier Series are said to be 'orthogonal'.) If in addition the stiffness **E** is the normal 'isotropic' array then the symmetric displacement components (u, v, w) and the antisymmetric components (u', v', w') also emerge separately. Thus a general problem can be solved as a series of separate smaller problems in each harmonic. This is computationally much more efficient than solving a single large problem. The displacements (u, v, w) and (u', v', w') are all functions of r and z only and are discretised in the same fashion as the axisymmetrical case. The resulting stiffness matrices are not too different from the single axisymmetric case ($n = 0$) containing

BRICKS, PLATES AND SHELLS

as they do simple additional terms like n^2, so assembly of each stiffness can be done without starting afresh for each harmonic. The details of this procedure are beyond the scope of this Primer and the interested reader should consult refs. 6 and 7. A good finite element system will automate the decomposition of general loading into harmonics (equation 6.4) for the user, and similarly at the end of the analysis will reverse the process and sum the stresses and displacements from each harmonic. The success of Fourier analysis relies on the necessary number of harmonics being small, in which case the number of nodal displacements in (u,v,w) will be considerably less than if (u,v,w) were discretised around the circumference also.

6.3 Plate Bending Elements

If a structure is thinwalled then we expect to be able to make reasonable approximations for the variation of strain or displacement through the thickness, and so eliminate one independent variable. Three dimensions becomes two. Chapter 5 was solely concerned with two dimensional membranes where all stress and displacements were uniform through the thickness. If thin flat plates are loaded normal to their plane, then bending must take place, and provided that displacements are small compared with the plate thickness there will be no induced stretching; that is membrane behaviour and bending behaviour are uncoupled. This is not true in curved shells in general.

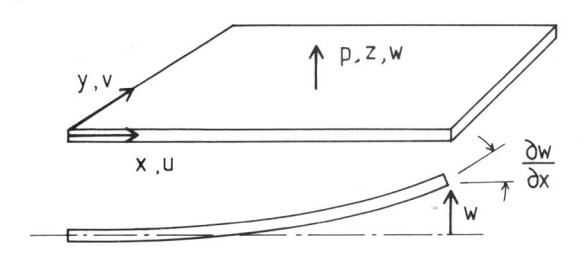

Figure 6.6 Plate bending.

It is tempting to analyse plate bending as a series of beams bending in two orthogonal directions. Indeed the first attempts by French analysts to derive the governing differential equations did just this – and consequently ignored completely the role of twisting moments and also the Poisson

A FINITE ELEMENT PRIMER

contractions. One of the very first attempts in 1941 to discretise plates also replaced them by a grillage (ref. 10). But we do not have to be so crude: we merely make the same kinematic assumptions that we did for beam sections (4.1) and write the two in-plane displacements components u and v in terms of the slope of the normal displacement (Figure 6.6).

Thus
$$u = -z\frac{\partial w}{\partial x} \quad \text{and} \quad v = -z\frac{\partial w}{\partial y}. \tag{6.7}$$

This leads to 'bending strains'

$$\varepsilon_{xx} = \frac{\partial u}{\partial x} = -z\frac{\partial^2 w}{\partial x^2}; \quad \varepsilon_{yy} = \frac{\partial v}{\partial y} = -z\frac{\partial^2 w}{\partial y^2}; \quad \varepsilon_{xy} = \frac{\partial u}{\partial y} + \frac{\partial v}{\partial x} = -2z\frac{\partial^2 w}{\partial x \partial y}. \tag{6.8}$$

Using only these assumptions, if we wished, the full plate bending equations could be derived in a manner identical to the beam theory of section 4.1 (see ref. 2). It is natural therefore to formulate a rectangular plate bending element by using the (Hermitian) cubic beam modes N_1, N_2, N_3 and N_4 of Figure 4.2, in both the x and y (or ζ_1 and ζ_2) directions. The nodal freedoms this time are the two rotations $\partial w/\partial x$, $\partial w/\partial y$, and the displacement w at each corner: 12 in all. The interpolation shape functions are simply products like $N_1(\zeta_1)N_1(\zeta_2)$, $N_1(\zeta_1)N_2(\zeta_2)$ and so on. Unfortunately when we examine the derivatives of these products we find that $\partial^2 w/\partial \zeta_1 \partial \zeta_2$ is zero at all four corner nodes, and equally seriously there is no constant component to this second derivative. As this controls the shear strain in equation (6.8), this violates the fundamental requirement that the element can represent all constant strain states. The performance of some systems' plate elements in torsion is not good, particularly for larger aspect ratios, and Figure 6.7 shows the results of one NAFEMS benchmark test.

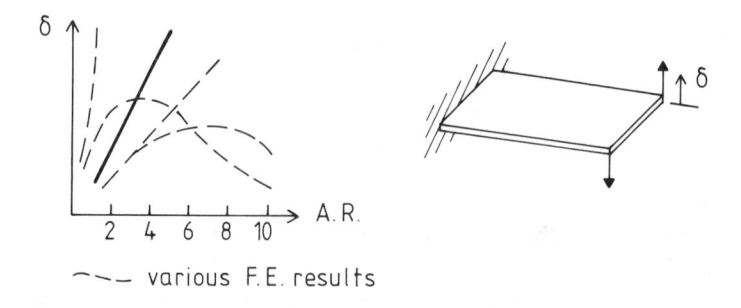

Figure 6.7 Plate Twisting and Aspect Ratio.

A compromise is to use the cubic Hermitian modes for the deflection shape functions but reduce the order to linear for the rotation shape functions. This element does have constant strain but is not conformal, and discontinuities in slope occur at interfaces. However, this element's results do converge as the size is decreased, and it is very convenient to use.

Other plate bending elements, both quad and triangular, avoid using products of separate one-dimensional shape functions, and expand the displacement in 12 term (or nine term) polynomials with varying degrees of success in achieving conformal elements, constant strain states, and symmetrical combinations of $x^n y^m$ to match the chosen degrees of freedom. A certain amount of element 'tuning' has also taken place over the years in some systems to remedy some of the deficiencies in these conventional plate elements based on equations (6.7) and (6.8).

One of the most popular bending formulations is now the Mindlin model which is based on an entirely different concept. It is straightforward to formulate and code. It resolves the problem of continuity between elements, and in one dimension is also a popular beam formulation as well. The starting point is to drop the main assumption of beam theory leading to (6.8) and which says that the (clockwise) rotations of the plate normally about the x and y directions are

$$\theta_x = \frac{\partial w}{\partial y}; \quad \text{and} \quad \theta_y = \frac{-\partial w}{\partial x}.$$

Instead we expand *separately* θ_x, θ_y, and w in terms of their nodal values. We still preserve the other beam-like assumption that normals through the plate thickness remain straight, and that strains ε_{zz} can be ignored. Figure 6.8 shows how the section now undergoes shear strain.

The displacement field is now

$$u = z\theta_y; \quad v = -z\theta_x; \quad w:$$

the direct strains are

$$\varepsilon_{xx} = \frac{\partial u}{\partial x} = z\frac{\partial \theta_y}{\partial x}; \quad \varepsilon_{yy} = \frac{\partial v}{\partial y} = -z\frac{\partial \theta_x}{\partial y}:$$

A FINITE ELEMENT PRIMER

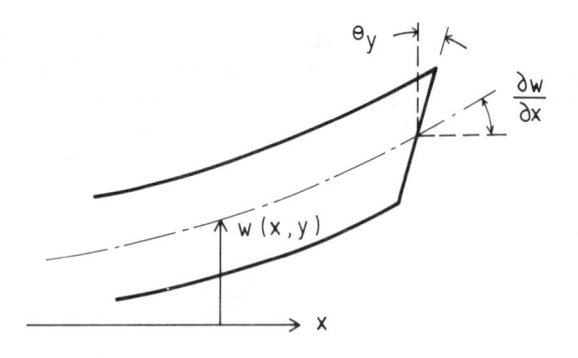

Figure 6.8 The Mindlin Model.

and the shear strain

$$\varepsilon_{xy} = \frac{\partial u}{\partial y} + \frac{\partial v}{\partial x} = z\left(\frac{\partial \theta_y}{\partial y} - \frac{\partial \theta_x}{\partial x}\right).$$

There are also through-thickness transverse shear strains which are now not zero, thus

$$\varepsilon_{zx} = \frac{\partial w}{\partial x} + \frac{\partial u}{\partial z} = \frac{\partial w}{\partial x} + \theta_y,$$

$$\varepsilon_{zy} = \frac{\partial w}{\partial y} + \frac{\partial v}{\partial x} = \frac{\partial w}{\partial y} - \theta_x. \quad (6.9)$$

A stiffness matrix can be found in the usual way and this element is generally quite well behaved, whether it is used as the common 8-noded quad form or higher-order. The bending strains in the 8-noded version are recovered accurately and the shear strains are as accurate if sampled at the reduced 2×2 Gauss points. In very thin plates the shear strains become relatively insignificant and near-zero values in (6.9) thus implies a dependent relationship between w and θ_x or θ_y. This form of ill-conditioning is usually avoided by using reduced integration or 'selective' integration by singling out the shear terms in $\mathbf{k}_g$ for special treatment. Nevertheless computers with a decent wordlength or double-precision arithmetic seem to handle adequately a standard 8-node element (using quadratic expansions for geometry as well as displacements w, θ_x and θ_y) and 'ordinary' reduced integration at the 2×2 Gauss points. Width/thickness ratios up to 50 seem perfectly safe. At the other extreme the Mindlin element is superior to classical plate elements if the plate is thick. The shear deformations become important when the thickness is greater than about a tenth of the plate width and this element represents the shear deformation directly without having to infer it as the derivative of the bending moment – an inaccurate approximation. The shear stress is admittedly constant through the thickness (instead of parabolic) so most systems introduce the correction factor of 5/6 when evaluating the contribution to the stiffness integral.

There is no perfect plate-bending element and we have not mentioned all the other possible formulations using mixtures of corner and midside freedoms in order to achieve near complete polynomials. Nor have we mentioned equilibrium formulations using simple expansions for moment distribution, or hybrid formulations using internal moments and edge displacements. No outright leader has yet emerged; and the same will be said for shell elements.

6.4 Shell Elements

Flat plates, depending on the loading, can behave as membranes with no bending strains, or alternatively as bending elements with no membrane action. The two are uncoupled unless the deflections are comparable with the plate thickness or larger. This is not true of curved shells where in general we should expect both membrane and bending to occur, just as in a two dimensional arch for example. In practice many shells are deliberately designed to act as membranes since any bending in a thin shell is likely to lead to high bending strains and an inefficient structure. The curvature of shells does indeed enable normal pressures to be resisted entirely by membrane action if the loading and the shape correspond in the correct fashion. A pure membrane shell is a very attractive structure which is incomparably superior to a flat plate or beam grillage in resisting normal loading. However most shells have to resist more than one loading action and a true membrane state is difficult to achieve everywhere, particularly as edge supports can probably be designed to resist only one loading case purely by membrane forces. It is usual to find that bending is unavoidable near concentrated loads, constrained boundaries, at sudden changes in load distribution or geometry – in fact any discontinuity usually means a bending zone in that region. Thus many shells consist of (perhaps large) areas where the efficient membrane action is the dominant shell behaviour, and other zones of bending where the membrane action is unable to simultaneously satisfy compatibility and equilibrium requirements.

A FINITE ELEMENT PRIMER

The finite element modeller should be able to guess fairly accurately where the likely stress concentrations are in order to be able to grade the mesh for maximum efficiency. The experienced engineer may even conspire to use pure membrane elements in favourable regions. In the case of flat membrane structures the stress concentrations are confined to the region of the discontinuity which produced them. They 'diffuse' away from this region. The same can happen in a curved shell membrane state, although it will not do so if the two local curvatures are in opposite senses (a saddle shape). In contrast if bending does occur in a thin shell then it is also favourably confined to a small region, but smaller than the flat plate diffusion zone, being of order $(rt)^{\frac{1}{2}}$ where 'r' is the local shell radius. The sketch in Figure 6.9 shows how bending stresses are confined to such zones in an axisymmetrical shell, and how the finite element mesh may be graded accordingly. Clearly much of the skill in analysing thin shells is in guessing the answer in the first place. The next step is to choose the right element.

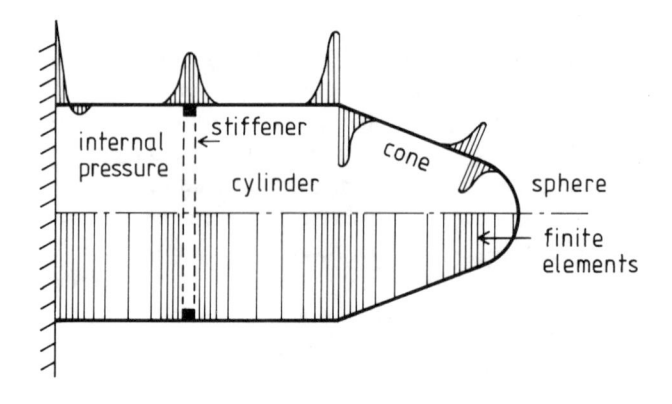

Figure 6.9 Bending and Membrane Zones.

A very crude and often extremely inaccurate way of analysing shells is to approximate the surface as a myriad of small flat 'facet' elements. The dangers are apparent immediately we think of special cases like the pressurised membrane solution. A pressurized cylinder for example has no bending stresses away from discontinuities, and if its shape is approximated by a series of flat quads there will be fictitious fold lines and fictitious bending stresses unless very small constant strain elements are used. The experienced analyst may be able to deduce the true nature of the solution but the novice is at risk, especially for configurations where the balance between membrane and bending action is not obvious.

There are other dangers in using flat quads or flat triangles. In a single flat element there is no coupling between bending and membrane actions. Thus if we add together in a single element the bending freedoms and membrane freedoms, there will be no term in the combined stiffness matrix which links them. The coupling occurs of course in the *assembled* shell stiffness where 'membrane displacement freedoms' at one fold line have 'bending displacement' components in the adjacent element. There are then problems with the three rotational freedoms at a node which are necessary to represent a general rotation. A flat membrane element has no in-plane rotational freedom about a normal through the node (at least those in current commercial use do not) and the rotational stiffness at a node will arise solely from the contiguous element's bending or twisting stiffness. But if adjacent elements are *co-planar* there will be rotational freedoms about plate normals which have no associated stiffness, and hence the global stiffness matrix will be singular. The offending freedom could of course be simply removed but there will be cases where elements are *nearly* co-planar, and the stiffness becomes ill-conditioned. Some systems deal with this by using fictitious spring stiffnesses to cope, whilst others dispense with corner nodes and use rotations along element edges – and pay the penalty of a larger band-width. The flat faceted shell is a dangerous simplification in the hands of beginners. More robust shell elements are advisable.

The difficulties of constructing thin shell elements are identical to those in plate bending, but are also compounded by the fact that the strain-displacement compatibility equations in thin shells are complicated and have been controversial. Some 4-sided and 3-sided finite elements, based on thin shell theory, do exist but they usually achieve completeness – or symmetric arrays of freedoms – by using both corner and midside nodes with different combinations of freedoms at each. Some also use higher order displacement freedoms like *curvature*, which is unnecessary for the PVD and which makes the element overstiff. Also curvature freedoms mean continuity of bending strains across element interfaces, and if the interface occurs at a change in thickness this will lead to incorrect discontinuities in moments. There is therefore a trend away from the classical shell formulation, and one towards treating shell elements as a special version of the three-dimensional brick elements, when one dimension becomes very small.

A FINITE ELEMENT PRIMER

One of the most straighforward isoparametric formulations is an extension of the Mindlin idea to curved elements.

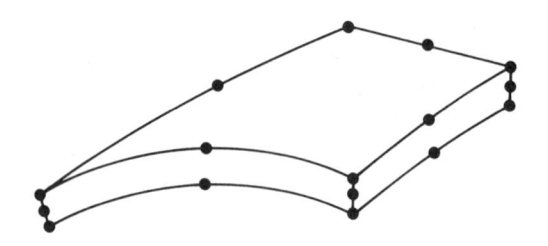

Figure 6.10 A 60 degree of freedom element.

Consider the general curvilinear brick of Figure 6.10. It would have several disadvantages if used to model thin shells. Firstly at the 20 nodes there are three displacement freedoms, making 60 in all. If all stress components are significant then **E** is 6×6 and **B** is 6×60, making the kernel of the stiffness matrix **B**^{t}**EB** extremely costly to compute. Secondly if the brick is thin its stiffness matrix becomes badly conditioned since the through-thickness stiffness becomes relatively large; or to take another view the displacement components normal to the middle surface become equal and so create a linear dependence in the total displacement vector **d**$_g$.

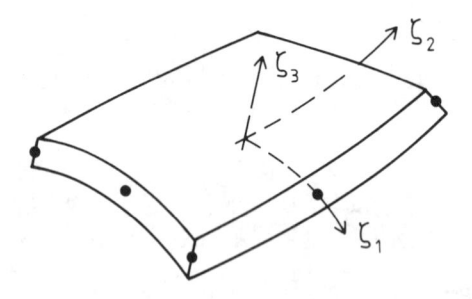

Figure 6.11 A 40 degree of freedom shell element.

The difficulties can be resolved by using the scheme of Figure 6.11 which deploys eight nodes around the edge of the middle surface $\zeta_3 = 0$ at the usual points ($\zeta_1, \zeta_2 = 0, +1, -1$). The middle surface geometry (and the in-plane displacements) are interpolated in the usual isoparametric fashion in terms of three coordinates (or three displacement components u, v, w). The coordinate ζ_3 and the element edges are normal to the middle surfaces and these are assumed to remain straight and unstrained. The displacement field can therefore be expanded in terms of the usual shape function **N** as

$$\mathbf{u} = \mathbf{N}[\mathbf{d}_1 + \tfrac{1}{2}t\zeta_3 \mathbf{C}\mathbf{d}_2]. \tag{6.10}$$

The displacements $\mathbf{d}_1$ (on $\zeta_3 = 0$) are the three u, v and w at the eight nodes, and those labelled $\mathbf{d}_2$ are the rotation components of the normal. The direction cosines of the normal **C** are obtained usually directly from the isoparametric description of the middle surface. Thus the degree of freedom of this element is $5 \times 8 = 40$. There are no through-thickness strains admitted and **E** must have the relevant components dropped, but (6.10) does allow transverse shears as in the Mindlin formulation for plates. For very thin shells some systems will therefore use 'selected' reduced integration on the shear terms in the stiffness matrix. It is proper to separate the membrane terms arising purely from $\mathbf{d}_1$ in (6.10) and use 2×2 integration on $\zeta_3 = 0$, but to use $2 \times 2 \times 2$ for the bending terms arising from $\mathbf{d}_2$; and there are other numerical tricks.

A related but different shell element is the Semi-Loof element of Irons (ref. 9) which has 32 degrees of freedom for a quad using three displacements at corners and midside nodes, and then normal rotations along each side at the two Gauss Points per side. This element can be cheaper to use than the specialised brick, but the oddly placed nodes and its incompatibility with other standard elements does not make it universally popular. In fact no element is. No shell element succeeds in all respects in passing the patch test, having zero rigid-body-motion strain components, being conformal and compatible with beam elements, coping with fold-lines, cheap to code, easy to use, and presenting convenient answers for stresses. It is one of the functions of NAFEMS to publicise the performance of shell elements either singly or in groups.

Finally, without going into details, it is possible and desirable to construct axisymmetrical thin shell elements in the same fashion as discussed for solids, and using Fourier expansions for non-symmetrical loading. Most systems have these elements which use either beam Hermitian (cubic) shape functions or else Mindlin type models. Those elements which use Hermitian shape functions, or even higher order, are uncontentiously derived from the full thin shell equations. It is probable that this route will

A FINITE ELEMENT PRIMER

in future be limited to axisymmetrical shells where the lines of principle curvature are clear and natural boundaries for both elements and the structure. For general shells where boundaries or intersections may be complex (see Figure 6.12) and where the shell surface itself may be far from a shell of revolution, it is likely that the degenerate brick element will be the main work horse.

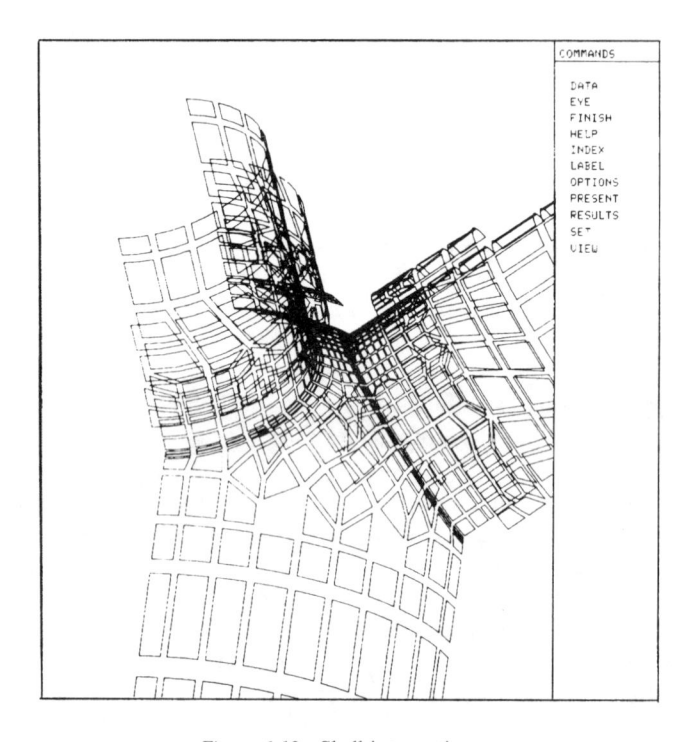

Figure 6.12 Shell intersections.

7. Mesh Specification

7.1 Introduction

In order to conduct a finite element analysis the structure must first be idealised into some form of mesh. The art of the successful application of the technique, so far as the user is concerned, lies in the combined choice of element types and associated mesh. As the method is approximate it is necessary for the user to have a good idea of the expected solution, together with an understanding of the consequences of the assumptions made within the element types to be used. This allows the effects of the approximation to be minimised within the solution. It generally requires skill and a good deal of experience to be able to define a mesh that will produce accurate answers at a reasonable cost for even a moderately complicated structure. Before a finite element mesh can be specified the problem that is to be analysed must first be identified. This requires the user to define four blocks of information:
a. The geometry.
b. The boundary conditions.
c. The loadings.
d. The required results.

It is a common mistake for the user of FE systems to think that only the first of these, the geometry specification, is needed before the mesh can be generated. The types of loadings and boundary conditions can affect the choice of element type or the locations of the node points but they are not usually especially critical for the mesh definition. The fourth information block, the required results, is the one that is most often overlooked at the mesh definition stage. However, this is the reason for carrying out the analysis and it is important that what is required is clearly defined and well understood before any computing is started. It can significantly affect the choice of element, the choice of the mesh and will always affect the cost of the analysis.

In order to conduct an FE analysis the structural continuum must be

idealised as a series of discrete finite elements. When specifying these elements four different sets of information must be given:
a. The nodal point coordinates.
b. The element topology (element node points and their interconnections).
c. The element geometrical properties; typically plate thickness or beam second moments of area.
d. The element material properties; typically Young's Modulus, Poisson's Ratio, density and coefficient of thermal expansion.

All finite element systems need this data for an analysis but every system has a different form of inputing data. At the least sophisticated level the user must give the coordinates for each node and the topology (nodal connections) explicitly for each element. Similarly the element geometry and material properties must be given in full. This is very tedious and error prone, but it does allow a completely general mesh to be specified under the complete control of the user. Most geometries require much less data for their definition than the full FE mesh does because the necessary element mesh density is defined by the rate of change of the strain energy in the structure. Almost invariably this changes in a more complex manner than the geometry, implying that various forms of automatic mesh generation facilities can be devised. At the most sophisticated level special pre-processor and mesh generation programs are available which are highly graphics-orientated and the geometry is constructed and meshed interactively. There are two extremes regarding the input data. At the lowest level the user does it all 'by hand' and, at the highest, it is done interactively by manipulating pictures. Within these two extremes there is a range of generation facilities available which are either an integral part of the analysis program or available as separate pre-processors.

7.2 Geometry Specification

If any form of automatic mesh generation is used it is common practice to define the structure's geometry first and, when this is done, to mesh it with suitable elements. The geometry is defined as either a series of lines, surfaces or volumes. It is also common practice to use lines to define a surface and then use surfaces to define volumes. Some programs also allow a direct specification of certain surface and solid shapes. Such a processor requires the user to specify the geometry as a series of regions that are either lines, surfaces, solids or combinations of these. In many cases regions correspond to the physical construction of the structure. It is useful to use such physically identifiable regions when the geometry is specified since this gives a logical grouping to the results when the output is printed. It also allows more than one person to build the geometry since each member of the analysis team can create a different component.

There are various difficulties that can be encountered at the geometrical specification stage of the problem. These vary according to the actual mesh generation procedures that are used. Lines can be defined in various ways. If they are straight or arcs of circles then the equation of the line can be given. More general lines are specified by some form of interpolation, which implies that the real geometry is approximated by a simpler curve. The accuracy of the geometrical specification is then dependent upon the interpolation functions that are used. Typically curves are approximated by parabolic arcs, or second or higher order splines. The higher the order of the approximating curve the more data the user must supply.

The basic problem with any form of automatic mesh generation is to ensure that elements which should be interconnected together are, but that they are not overconnected. It is usually possible for the analyst to specify the specific nodal connections but any mesh generation program should do this automatically. One way of achieving this is for the program to connect nodes together that are within a pre-defined radius of each other. The system (or the user) then has to choose a suitable small radius. If it is too small, nodes that should be joined will be left underconnected, but if it is too large then nodes which should be separate will be joined. This is especially a problem with structures that have a high aspect ratio or where the same idealisation contains both a very fine and a very coarse mesh. A rather more subtle connection error arises where two adjacent edges are defined by curves of different orders. To illustrate the problem consider the generation of the geometry of Figure 7.1, where the common line is a circular arc.

The dotted lines show a typical element mesh. If the common circular arc of the inner region is defined by a parabolic approximation and the adjacent edge of the outer region is also fitted by a curve of the same order, then, because of the different lengths of the two arcs, the two approximations will not give rise to the same equation and intermediate points along the common edge will not necessarily be connected correctly. This problem is especially difficult to detect for three dimensional solid meshes where points on adjacent faces can be left free. Obviously the existence of spurious cracks in the model that arise from such a fault can cause serious errors in the final solution. Any such spurious cracks should

A FINITE ELEMENT PRIMER

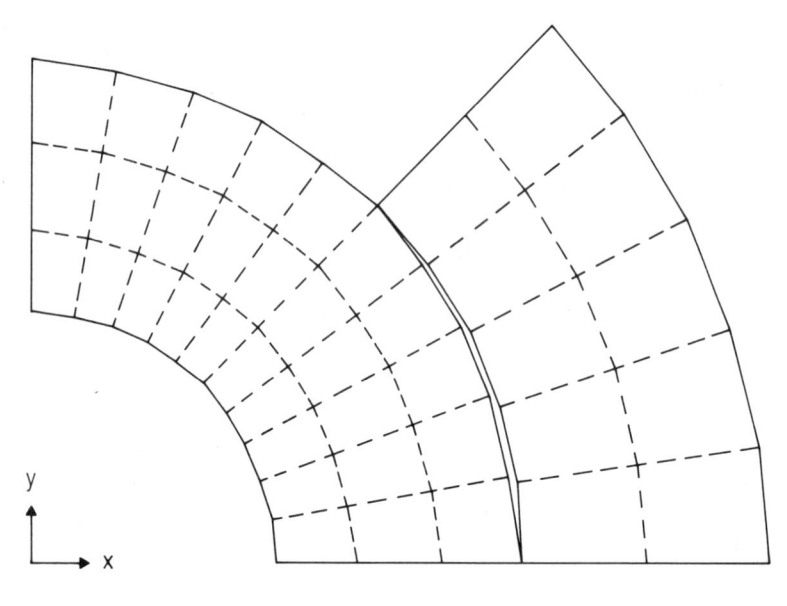

Figure 7.1

be detected and corrected as soon as possible within the analysis. In passing it is worth noting that, although the three points defining the parabolic approximation to the circular arc are equally spaced, the generated mesh does not have a uniform spacing around the arc. A mesh generator that is based upon the isoparametric element shape functions only generates equally spaced nodes along regions with straight edges.

The approximation of a real curve by some lower order one can also lead to geometrical errors. When two lines are joined end to end then the slopes will be discontinuous, as shown in Figure 7.2. This can lead to an artificial geometrical stress concentration. It is also possible for the interpolation approximation to cause the highest curvature on a line to be slightly away from a stress concentration. This has the effect of 'smearing' the stress concentration within the analysis. The higher the order of the interpolation function that is used to define the lines then the less likely is it that slope discontinuity will present a problem. It only becomes significant where the geometry has a high rate of change of curvature. A relatively low order interpolation function can still match segments of the curve to high accuracy but the slope discontinuities can introduce psuedo stress concentrations giving a non-smooth stress distribution. From a purely geometric point of view this problem is overcome by using spline

MESH SPECIFICATION

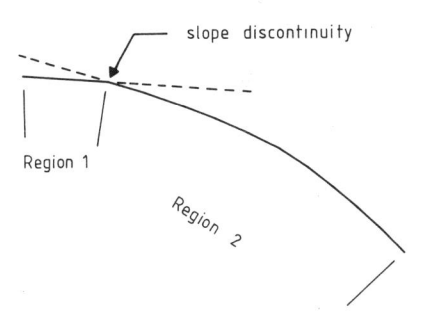

Figure 7.2

curves, where end slopes are matched between the splines. However, the error can still exist within the finite element mesh since most element shape functions are defined by some low order polynomial and there is no slope continuity between the elements. An extreme example of this is the modelling of curved shells by a series of flat elements. The resulting faceted geometry can lead to oscillations (sometimes quite violent) in the computed results.

The user must take care when using a mesh generator with the combination of higher order elements with mid-side nodes together with a variable mesh density. If the generator uses a simple mapping then the mid-side nodes of the element will not be generated in their correct mid length position. This can be corrected for where the edges are straight but it requires the generator to do something rather more sophisticated than a simple mapping for curved edges.

Probably the most difficult aspect of the geometry specification is related to the intersection of solids and surfaces. The problem usually resolves itself within the FE mesh generation to the definition of lines of intersection. General purpose mesh generation programs contain facilities for finding and blending such intersections, but these should be used with care. Where the geometry of the two intersecting surfaces are significantly different then a stress concentration will arise along the intersection line. If arbitrary blending facilities are used the magnitude of this concentration can be significantly affected by the blend geometry chosen by the program. If the stresses around the area of the intersection are important then the user should confirm that the blending used does in fact correspond to the real geometry.

A FINITE ELEMENT PRIMER

7.3 Mesh Generation

Once the geometry has been defined then it must be meshed with elements and most FE programs give the user a wide range of element types. The choice of which of these to employ will be restricted by the geometry definition that has been used. If this is in terms of surfaces then membrane or shell elements must be used, but if a three dimensional solid geometry has been constructed then it is most natural to use solid elements. The majority of FE programs in current use, and all of the most commonly used elements, are based upon the displacement method and this allows families of element types to be defined. Typically a family contains elements that use either linear, parabolic or cubic interpolations. Current practice indicates that second order (parabolic) elements give the best compromise between accuracy and efficiency for general use.

The choice of the mesh density is dictated by the element type that is to be used and the expected stress distribution throughout the structure. Any finite element is only approximate and the inherent approximations must be considered when the mesh is being generated. For the displacement method they arise in the satisfaction of the equilibrium conditions, the compatibility conditions are usually satisfied exactly within the element. The user should ensure that compatibility is then satisfied throughout the structure. If this is done then it can be guaranteed that the structural displacements will converge monotonically as the element mesh is refined. If compatibility is not satisfied throughout the structure this convergence statement cannot be made and the results can oscillate about the final solution as the mesh is refined or, worse, they can converge to the wrong answer. Some elements, especially plates and general shells, do not satisfy compatibility internally. Under-integrating the element stiffness matrices by means of reduced integration also has the effect of violating compatibility within the element.

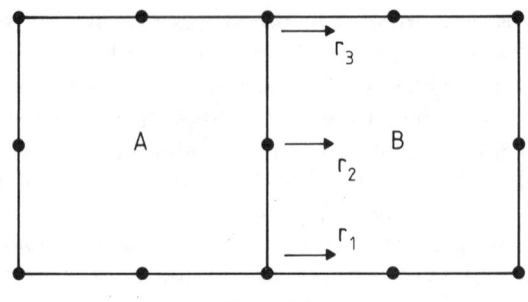

Figure 7.3

Complete compatibility will only be satisfied if elements with the same interpolation function, at least along the common edge, are used. Figure 7.3 shows two 8-node quadrilaterals that are interconnected. Along the common edge the horizontal displacements are r_1, r_2 and r_3. The horizontal displacement at any point along the common edge of element A is fully defined by these and the assumed element interpolation function. Element B uses the same interpolation and has the same displacement and hence the displacement at any point along its common edge will be exactly the same as for A and there is complete compatibility along the edge. The assembly process joins not just nodes but also complete edges, provided that each element has the same edge interpolation function, giving full satisfaction of compatibility throughout the mesh.

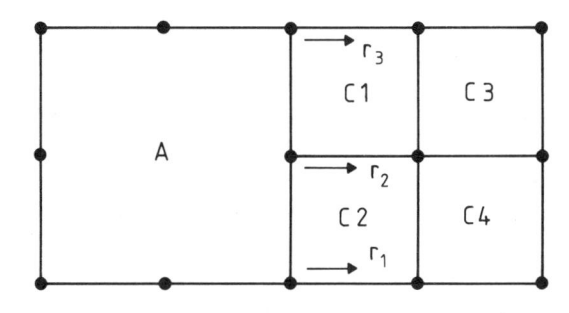

Figure 7.4

Consider now the mesh of Figure 7.4, where element A is as before but element B has been replaced by two elements C1 and C2, both with a linear interpolation of displacements. The horizontal displacement of element A still follows the same parabolic shape of the interpolation function but the displacements for elements C1 and C2 are interpolated linearly and, although the nodal displacements are matched, the intermediate horizontal displacements are not. There are then gaps between the nodes of the elements and the calculated stress distributions will try to represent the behaviour of these gaps. The resulting errors depends upon the loading and the position in the mesh where the interface occurs. They can have a significant effect upon the calculated stress distribution and, although most finite element programs allow the user to construct such a mesh, it is a procedure that should be avoided since it can lead to unreliable answers. It should certainly not be used as everyday practice or by inexperienced users.

A FINITE ELEMENT PRIMER

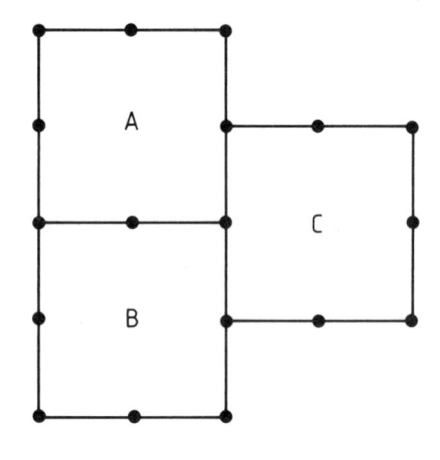

Figure 7.5

Continuity can be violated even where elements of the same type are used. In Figure 7.5 all three elements A, B and C are of the same type but, because the corner nodes are connected to mid side nodes, the interpolation functions along the common interface are not the same.

It is possible to connect dissimilar element types provided that common interpolations are used along the interface. Some examples of this are shown in Figure 7.6. In Figure 7.6a a second order triangular element is connected to a second order quadrilateral. In Figure 7.6b a special transition element is used between first and second order elements. The

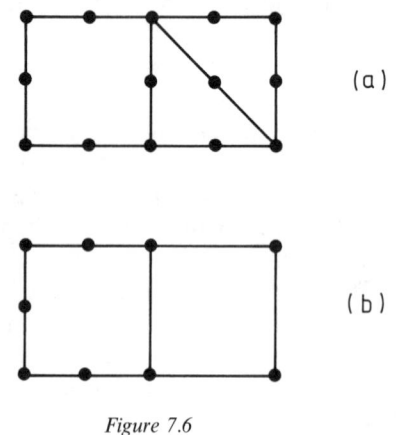

Figure 7.6

shape function of the transition element is such that the interpolation is parabolic along the edge that connects to the second order quad and is linear on the edge that connects to the first order quad.

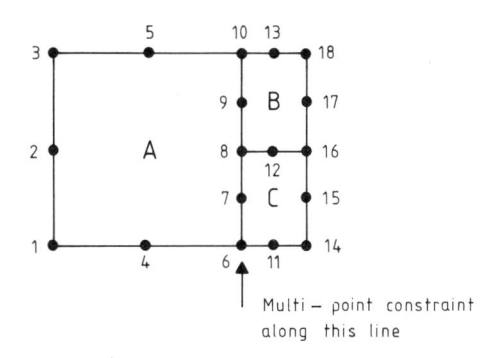

Figure 7.7

Some programs have facilities for applying multi-point (or generalised) constraints and these can be used to allow abrupt changes in element geometries whilst still retaining compatibility. Consider the mesh shown in Figure 7.7. As it stands there is not complete compatibility between element A and elements B and C. Nodes 7 and 9 do not occur on element A and therefore constitute a crack. However, multi-point constraints can be used to enforce compatibility by constraining the displacements at nodes 7 and 9 to be defined by the displacements at nodes 6, 8 and 10. This is discussed in detail in chapter 12.

Compatibility can only be enforced by such constraint equations if the elements that are to be constrained have the same interpolation functions along the interface boundary. In Figure 7.7 elements A, B and C all have the same shape functions and, in particular, the interpolation function along the interface edge of element B can match the interpolation of the portion of the interpolation function of element A.

Multi-point constraints can also be used to match elements with dissimilar shape functions. In Figure 7.8 element A has a linear interpolation and element B is parabolic. Constraint equations defined by the linear interpolation of element A can then be used to define a constraint for the mid edge displacement at node 4 for element B so that this end only distorts in a linear manner. Constraint equations cannot be used to enforce compatibility between dissimilar elements for the mesh of the form

A FINITE ELEMENT PRIMER

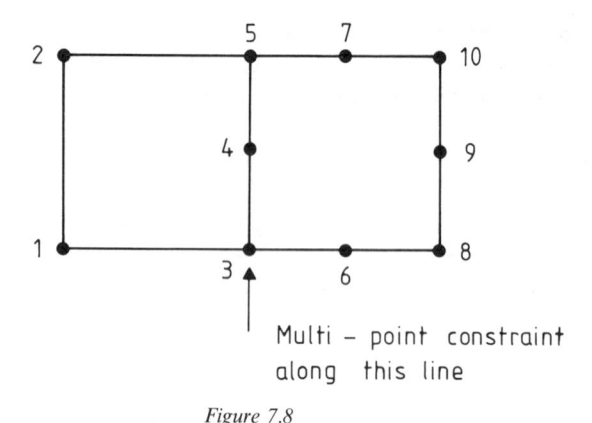

Figure 7.8 — Multi – point constraint along this line

shown in Figure 7.4 where there is a change in element density in addition to a change in element interpolation.

7.4 Mesh Density

The art of using the finite element method lies in choosing the correct mesh density required to solve a problem. If the mesh is too coarse then the inherent element approximations will not allow a correct solution to be obtained. Alternatively, if the mesh is too fine the cost of the anlysis can be out of proportion to the results obtained. In order to define a relevant mesh then some idea of the stress distribution is required, that is, if the answer is known then a good mesh can be defined! Fortunately this extreme is not necessary but the user still needs to know which areas of the structure have high stress gradients. It is also necessary for the user to know the element behaviour and the approximations used in its formulation. A fine mesh is required where there are high rates of change of stress (and strain) and a coarse mesh can be used in areas of reasonably constant stress. This still begs the question as to what constitutes a fine mesh. A linear displacement element requires a finer mesh than a parabolic one which, in turn, requires a finer mesh than a cubic element.

High rates of change of stress will occur whenever there is any form of discontinuity. This can occur in the geometry, in the loading or in the material properties. Generally a finer mesh is required in such regions than elsewhere in the structure. The equations of elasticity are such that any form of discontinuity only produces a local disturbance to the stress field and the effect of the discontinuity dies away with distance, producing a local stress concentration (St Venant's Principle). There will be a diffusion or die-away length associated with the local stress discontinuity and it is this length that defines the changes required in the element mesh density.

The fact that the effect of a stress concentration dies away means that a graduated mesh can be used over the die-away length associated with discontinuity. The mesh can be made much coarser away from the discontinuity. On some occasions it is not the purpose of the analysis to find the magnitude of the stress concentration associated with a particular discontinuity but the stress in some other region is required. Provided that the region of interest is sufficiently far away from the discontinuity (beyond the die-away length) a coarse mesh can be used in the region of the discontinuity, with the user recognizing that accurate stresses will not be recovered here. The stresses can still be very accurate in the actual region of interest provided that the overall stiffness of the coarse region of the mesh is correct. In effect the user is specifying a coarse model for the parts of the structure that are not of direct interest such that it provides the correct boundary conditions and load transmission paths to the regions that are of interest. However, if this is done then it should be noted in any documentation associated with the model that the stress results in such areas should not be used for assessment purposes.

It follows from the above discussion that any user of a finite element program must have some knowledge of structural analysis and the manner in which a given structure behaves. He must be able to identify regions of stress concentration and be able to estimate die-away lengths associated with a given form of discontinuity. It is not possible to give any simple rules for these lengths because different geometrical forms and different loading systems have different lengths associated with them. It is also possible to have a series of stress concentrations superimposed upon each other. Typically where a cylinder meets a heavy flange in a shell there will be a local stress concentration that arises from matching overall displacements and rotations at the junction. Superimposed on this there can be an even more local effect arising from a sharp corner where the cylinder and the flange meet. In such cases it is possible to devise a mesh that will find the sum total of both stress concentrations, or a coarser mesh that finds the stresses due to the matching overall compatibility but not those due to the sharp corner. Finally, a yet coarser mesh that only gives the displacement behaviour can be set up. Before a mesh is chosen the user must decide what he is trying to achieve by the analysis and choose the elements and the mesh accordingly.

A FINITE ELEMENT PRIMER

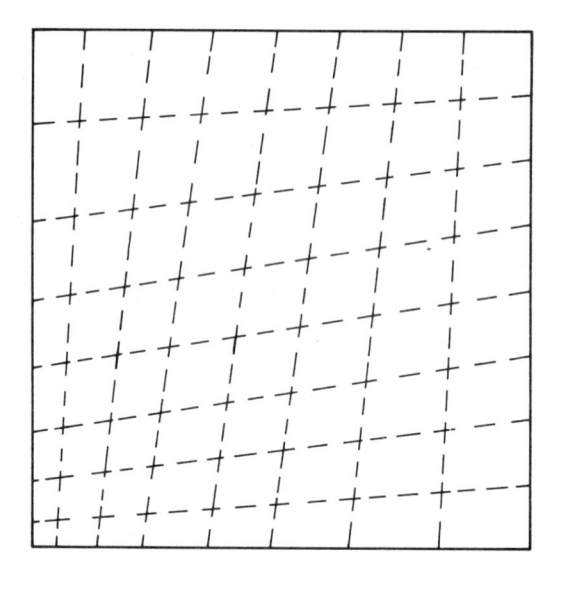

Figure 7.9

Once the region requiring the finest mesh and the actual size of the mesh at this location has been decided upon then the user is faced with various choices regarding the mesh for the rest of the structure. The simplest is to have a uniform mesh of the smallest size elements all over the structure. This is rarely done because of the analysis cost that such a strategy can incur, especially for three dimensional structures. If a uniform mesh is not used then a variable mesh density must be defined. One of the main advantages of the finite element method over other numerical solution techniques is that it is inherent in all FE programs that variable density meshes can be used. There are two basic methods for altering the mesh density. In the first the element sizes are varied in some gradual fashion, as illustrated in Figure 7.9. Here there is a fine mesh in the bottom left hand corner of the region and a coarse one in the top right hand corner. This is usually done automatically as a part of the mesh generation process. Note that the number of elements along opposite edges is not changed, only the nodal spacing is different. In the second method of varying the mesh density the number of elements along opposite edges of a region can be different. Various methods for doing this are shown in Figure 7.10. If triangular elements alone or combinations of triangular and quadrilateral elements are used then any degree of mesh refinement can be accommodated. It is possible to refine the mesh using quadrilateral elements alone but there is then some small restrictions upon the possible

MESH SPECIFICATION

changes in mesh density from side to side of the region that is being meshed. For arbitrary changes in mesh density then the mesh must be specified by the user, at least for the irregular portions of the mesh. Some programs, especially specific mesh generation pre-processors, contain various forms of automatic mesh refinement as macro operations. It is possible to devise similar schemes for changing mesh density when using three dimensional solid elements with combinations of hexahedral, tetrahedral and wedge elements but the process is very much more difficult to visualise in this case. Mesh generation programs tend to contain fewer automatic variable mesh generation facilities for three dimensional geometries.

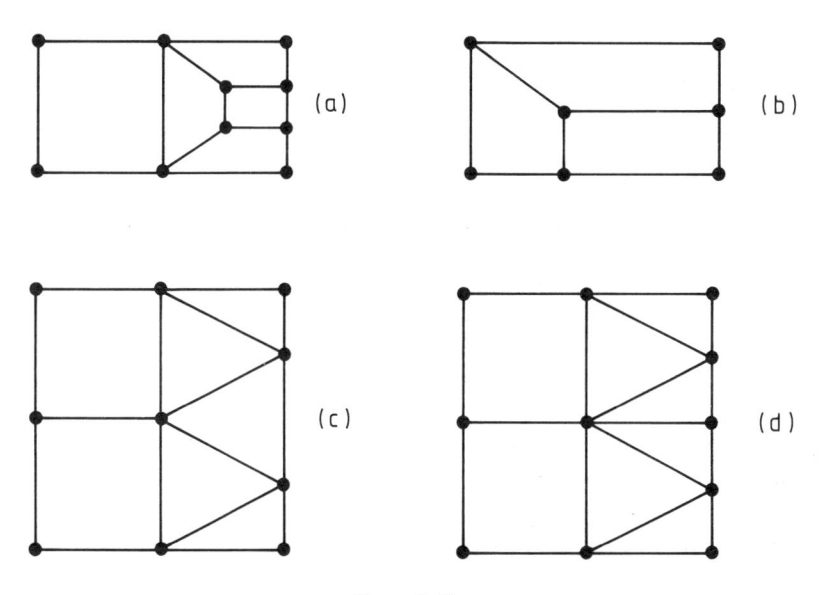

Figure 7.10

It is not difficult to control element shapes where the geometry and the mesh is regular but less control is available for irregular meshes or geometries. There are limits to the degree of distortion that an element type can be made to undergo. Every element is defined in terms of the basic shape of a parent element, typically the basic shape for a quadrilateral is square, for a triangle an isosceles triangle, and for a hexahedron a cube. When elements are distorted from the shape of the parent they are found to be less accurate. As the distortion is increased the greater the error in the element behaviour. In setting up a mesh the

97

user should attempt to keep the elements as near to the basic shape of the parent as possible. The distortion that an element will tolerate depends upon the element type and, to a lesser extent, the loading. There are four possible forms of element distortion, these being:
a. Aspect ratio distortion (elongation of the element).
b. Angular distortion of the element, where any included angle between edges approaches either 0 or 180 degrees (skew and taper).
c. Volumetric distortion of the element, where the mapping between the real element space and the non-dimensional basis space is such that the volume transformation may tend to zero at some point.
d. Mid node position distortion. This only occurs with high order elements.

The various forms of distortion are illustrated for a quadrilateral element in Figure 7.11. All of these can be quantified by means of the element Jacobian at any point. The details of this are discussed in chapter 8. Various simple rules of thumb can be stated, which if followed at the mesh generation stage will usually mean that the elements are not overly distorted. Obviously, for the aspect ratio any one side of an element must not be significantly longer than the shortest side. With angular distortion all included angles should be kept as near equal as possible, typically for a quadrilateral element the included angles should be close to 90 degrees. Volumetric distortion occurs where the element is so distorted that it is turning back on itself and it is usually symptomatic of either aspect ratio or angular distortion but it can arise with other peculiar forms of mapping from the basis to the real space. The distortion due to incorrect positioning of the mid side node is interesting in that the element can appear to be well formed at first sight, but if the mid side node is not placed centrally then there is an effective distortion due to the mapping from the basis space to the real space. For a quadratic interpolation then positioning the mid side node at the quarter chord point leads to an element with a singularity in its stress distribution. This is discussed in detail in section 5.4. Higher order elements have a similar behaviour but with different positioning of the mid side nodes. When meshing any real structure some degree of element distortion must be used in the mesh. The permissible degree of distortion depends upon the element type and the stress field that the element is subjected to. Generally the higher the order of the element or the smaller the stress gradients across it then the more it can be distorted from the parent shape. It is usually considered to be the users responsibility to carry out checks to decide what are the maximum allowable distortions for an element for his own special application, although some programs do make checks upon this and a warning is triggered if a preset level of distortion is exceeded.

MESH SPECIFICATION

Aspect ratio

Angular distortion (skew)

Angular distortion (taper)

Curvature distortion

Mid-side node position

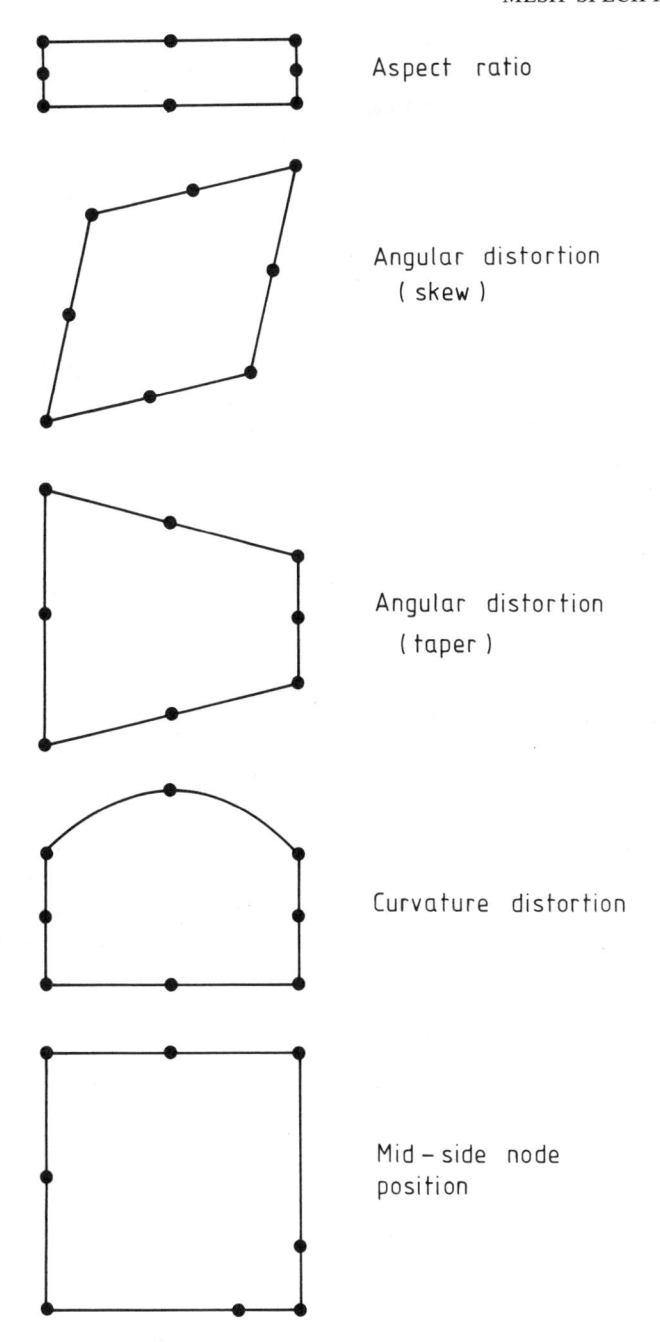

Figure 7.11

All of the element distortions can be related to the Jacobian and interpreted as some function of the terms in the Jacobian going to zero. If this function goes to zero within the element then there is a numerical singularity at this point that arises solely from the mapping function. Since the singularity is not real it can lead to large errors in the stress results. It will have less effect upon the computed displacements. If the function does not go to zero within the element it can be extrapolated to find where it is zero and, so far as the element is concerned, there is a stress singularity at that point. If it is a significant distance from the element (greater than the typical die-away length for a stress concentration) then it will have little effect upon the accuracy of the computed stresses. There are occasions where this singular behaviour is used to advantage to models cracks but care must be taken to ensure that the form of the singularity that arises from the mapping corresponds to the physical singularity.

So far as it is possible a user should try to generate a mesh such that there is no large changes in stiffness across element boundaries. For uniform materials and continuous geometries this means that similar sized elements should be used on either side of the boundary. As a rule of thumb the volume of one element should not be more than two to four times the volume of any immediately adjoining element, although this depends upon the elements actually used and the stress gradients across the element. If there are abrupt change in the material properties across element boundaries then the user should try to make the total stiffness for the elements on either side of the boundary about the same. This is not always easy to do and, where possible, the variation in the material properties should be smeared across a band of elements since this gives a gradual change in the stiffness and leads to more reliable results. As a general rule the more gradual that any transition of material properties, loadings or geometry can be made then the better the results will be. This applies equally for changes in mesh density which should be gradual rather than abrupt.

Problems often arise in the modelling idealisation where physical discontinuities in the structure occurs, typically in the modelling of joints and other connections. It would appear at first sight that the FE method is ideally suited to solving such problems since elements are interconnected at nodes which can be made to coincide with connectors. Unfortunately it is not so simple as this, since the FE method has the effect of smearing the structural behaviour over the element rather than concentrating it at nodes. The method itself only guarantees mean square convergence over the element, not necessarily convergence at individual nodal or other points. The modelling of joints then requires some form of approximation and the facilities that are available within individual programs have to be considered in order to decide what idealisations are most suitable. If some freedoms have to be interconnected to form the joint then this is best done by means of multi-point constraints to join the freedoms together in some specified pattern. Some elements have rigid offset facilities built into their formulation and these can be used to alleviate modelling problems associated with connecting beams and shells. They are used to connect the neutral axis of a beam to the centre line of a plate. If these facilities are not available, or for some other forms of interconnections, then degrees of freedom can be connected using fictitious very stiff elements. However, this is a very dangerous practice and has been responsible for many difficulties in using the finite element method in practice. It should be avoided wherever possible. The user has to be very careful about the choice of the property values used to define such fictitious elements. If they are not high enough the constraints will not be satisfied, but if they are too high then the resulting equations will be ill-conditioned and errors can arise in the solution process. The user should verify that the same results are obtained when the properties are changed by an order of magnitude about the value used.

Problems can also arise in modelling contact or other gap types of problems. This is often done using pin-jointed bar elements to connect freedoms across a gap and giving these either a very high stiffness if the gap is closed or a very low stiffness if the gap is open. The program, or the user if it is being done manually, then iterates adjusting the stiffness of the gap elements to be one or the other of these high or low values until convergence is achieved. Some programs have special gap elements which work in this way and are little more than bar elements with automatic iteration facilities. The use of such bar and gap elements is restricted to simple linear interpolation elements and, unless considerable sophistication is employed, they should not be used with higher order elements. The reason why they are unsatisfactory with higher order elements stems from the fact that the elements only give mean square and not pointwise convergence. The kinematically equivalent forces for a uniform normal stress along the edge of a high order element show considerable variation from node to node. The forces in bar and gap elements show exactly the same large variation and the fact that the force in a bar is positive does not necessarily mean that there is a tensile stress on the element at the node where the bar connects. This makes the use of simple gap elements very unreliable when used in conjunction with the higher order elements.

If the user has any doubts as to the reliability of such elements then a simple model should be constructed to test them. A form of patch test is probably the best, with the patch having a combination of elements that are being used in the analysis and the loading is such that the model should be in a state of uniform stress. The same argument applies also the the calculation of reaction forces at support points when high order elements are used. The size of the reaction at a point need not give any indication of the support force that will actually be felt at that point. At best the average reaction force along the edge of an element is all that can be used. This illustrates the mean square smearing effect that is inherent in the FE method.

7.5 Choice of the Element Type

Most general purpose FE programs provide the user with a wide range of choice as to the element types that he can use. It is up to him to choose from this library of elements the ones that are appropriate to solve his particular problem. Although, for any program, there are recommended elements which should be used, the possible choice is still large. The elements used for a given analysis depend upon the method by which the structure carries the loadings. Element behaviour can be classified into one of five categories:

a. Membrane – where the element carries only load in its plane and has no stiffness normal to the plane.
b. Bending – where the element carries loads normal to the plane in which it lies.
c. General shell – which is a combination of classes a. and b.
d. Solid – for analysing a three dimensional continuum.
e. Axisymmetric – where the geometry is a body of revolution.

In a well designed structure there are specific load transmission paths that carry the applied forces to the equilibrating support points and these define which elements should be used. For example, in stressed skin aircraft construction any concentrated forces, such as control surface loads, are passed to the skin through relatively rigid diffusion members. The skin itself is supported by light frames which maintain the cross-sectional shape of the structure. Although the skin is thin it is supported against bending. The skin acts as a membrane, carrying only stresses in its own plane allowing the structure to be modelled with membrane elements. Similar arguments can be used when considering how to idealise other structures, especially those which have been designed to carry a specific form of loading. Obviously, to achieve this the user must have a full understanding of the behaviour of the structure. In theory it is possible to analyse a structure without considering how it works but this can be very dangerous. Invariably a cheaper analysis can be conducted if the actual behaviour is modelled and often the numerical stability of the solution process is also better. For the airframe construction discussed above the use of shell elements doubles the number of degrees of freedom since nodal rotations have to be included. The conditioning of the resulting set of simultaneous equations will be worse than for a membrane model because of the very low bending stiffness of the skin. It is possible for the more detailed and expensive model to produce worse answers than the simpler one. These effects become magnified when considering the use of solid element.

Inexperienced users tend towards overkill when setting up an FE model, wanting to use three dimensional solid elements to model everything. Although solid elements can, in theory, be used to model any structure this is not a good idea unless the structure really is a three dimensional continuum. The cost of any 3-D solid model analysis is likely to be an order of magnitude more expensive than the equivalent 2-D one. It is more expensive both in money (or computer resource) terms and the increase in man effort required to set up and debug a 3-D mesh. A less obvious reason for not using solid elements where the structure is not a true solid is concerned with the numerical stability of the set of equations that have to be solved. A three dimensional model will have considerably more equations than a 2-D one but this increase is not the direct cause of the ill-conditioning. If the structure is thin in one direction then the stiffness in this direction is found from the difference of very similar numbers. On computers with a fixed word length this becomes progressively less accurate as the structure becomes thinner making the equations become ill-conditioned. In the limit the fixed word length will make them singular. Solid elements are not good at modelling thin shells. For a shell element the through-thickness effects are idealised out of the element formulation and they do not have the same conditioning problem, although they do have their own problems if they are not used in the correct context.

The various modelling possibilities that are available to the user can be illustrated by considering the analysis of a simple cylinder. The possible alternative element types, in order of increasing analysis costs are:
a. An axisymmetric thin shell.
b. An axisymmetric thick shell.
c. A general thin shell.
d. A general solid.

A FINITE ELEMENT PRIMER

AXISYMMETRIC THIN SHELL

AXISYMMETRIC THICK SHELL

GENERAL THIN SHELL

GENERAL SOLID

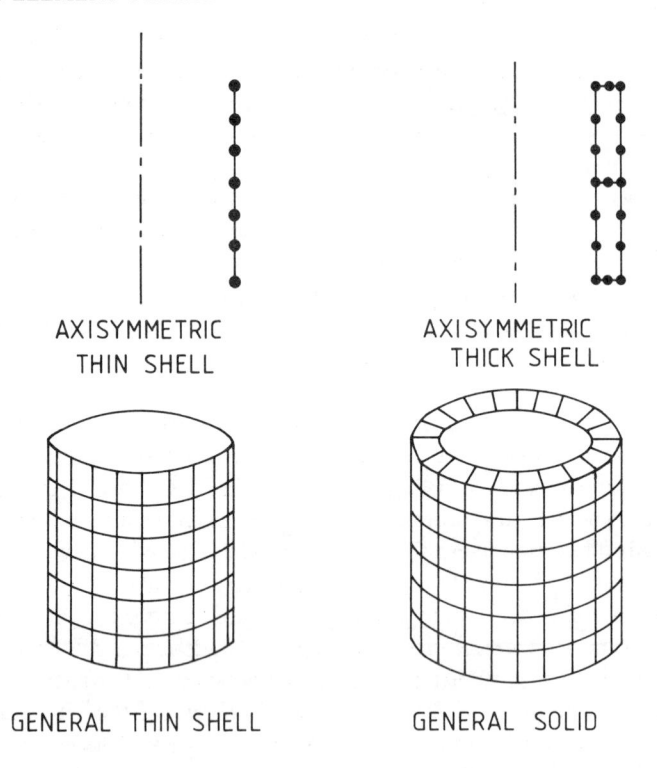

Figure 7.12

These options are shown in Figure 7.12. If the cylinder is thin, that is the radius to thickness ratio is greater than 10 say, and the loads do not vary violently around the circumference of the cylinder then option 1, the axisymmetric thin shell model will give a very accurate and cheap analysis that can be performed and assessed rapidly. If the cylinder is not thin or if the loads or geometry are such that the stresses vary in other than a constant or linear manner through the wall thickness then an axisymmetric thick shell element can be used. This will require at least twice as many nodes as the thin shell since there must be a line of nodes at least on the inner and outer faces of the cylinder walls rather than just along the mid-thickness line. The thick shell model will be less well conditioned numerically than the thin shell model. The displacements in the radial direction of the cylinder define both the hoop and the through thickness stiffness of the shell. As the radius to thickness ratio is increased then, relatively, the hoop stiffness falls and the through thickness stiffness increases causing ill-conditioning as the shell becomes thinner.

104

MESH SPECIFICATION

The general thin shell model has this possible source of error eliminated at the element formulation level. It will require a considerably more expensive computation than the axisymmetric thin shell model because it has many more nodes. It is also usually more expensive than the axisymmetric thick shell because the equations of the model have a much larger bandwidth. It is superior to both axisymmetric models in that it allows easier modelling of general loadings and it also allows both the geometry and the material properties to vary around the circumference of the shell. The use of thin shells in this form can introduce geometrical modelling errors. Some FE systems only have flat thin shell elements and the cylindrical surface must be modelled as a series of flat facets. The discontinuity in the geometry that this introduces can give rise to oscillations in stresses and displacements around the circumference of the cylinder that can only be mitigated by making the included angle subtended by the element small. This then increases the number of elements that are required around the circumference of the cylinder, slowing the convergence of the solution as the mesh density is increased and increasing the cost of the analysis. The same applies to higher order elements to some degree since, although these can be curved, they do not fit a cylindrical surface exactly. Again as the subtended angle is increased there is a slope discontinuity in the modelled geometry, although this is much less of a problem than the flat facet limitation.

If a solid element is used to model the cylinder then all of these problems are compounded. For a simple cylinder with a moderate radius to thickness ratio the axisymmetric thin shell element will deliver a more accurate solution for the overall behaviour of the cylinder at a fraction of the cost of a solid element solution. The cost that is paid for an analysis, in computer resource or man hours in the model preparation, is not necessarily proportional to the validity of the results.

7.6 Testing Element and Mesh Suitability

With the many possible alternative options of element types, coupled with the wide range of different forms of structures that have to be analysed, it is impossible to give a simple set of rules for using the finite element process. Instead the responsibility is shifted to the user, requiring him to have a good knowledge of the behaviour of structures and the behaviour of the elements that are being used. It is not possible to give rules for ostensibly similar element types since the formulation of elements can vary within different programs. As an example some systems contain 6-noded triangular elements which are formed by defining the explicit shape functions, whilst other systems collapse a line of nodes of an 8-noded quadrilateral to form the triangle. Although the two elements so formed are geometrically identical they behave differently for a stress analysis. This requires the user to give different considerations to the choice of meshes, depending upon the actual formulation of an element used by the system. It is also possible to have elements in different systems that have identical shape functions but which use a different number of numerical integration points in forming the stiffness matrices. This can lead to the elements having a different behaviour. All of these considerations serve to show that the user must establish for himself the behaviour of individual elements in particular systems. It is not possible to follow blindly any generally accepted rules since the particular implementation of even a 'standard' element can alter its behaviour. It also follows that what is a suitable mesh for one FE system need not be suitable for another, even with nominally identical elements. One of the aims of NAFEMS is to provide the finite element community with a source of information and standard tests that can be used to establish the behaviour of any element in any system.

In order to establish the mesh density for a given analysis the user should run a series of simple test cases before the actual analysis is attempted. As a first step the requirements for the convergence of the finite element method should be considered. The essential idea of the technique is that the approximations incorporated into the element interpolation functions become closer to the exact solution as the mesh is refined. In the limit the element must be able to undergo strain free rigid body movements and be able to reproduce constant stress conditions. If the element does not allow strain free rigid body motions then it is effectively mounted on elastic foundations and such an assembly will not represent the true behaviour of the structure. The problem of not having strain free rigid body motions can arise for elements that have been distorted from the parent geometry or for elements that contain some form of internal constraint that goes to ground. One reason for the success of the isoparametric element formulation is that it can be shown that, if the parent shape of such an element has strain-free rigid body movements then any distorted form of the element also has this property. Non-isoparametric elements tend to be susceptible to this form of error and the user should investigate the behaviour of distorted non-isoparametric elements very carefully before they are used. There are some elements that contain internal 'bubble' functions that considerably improves their behaviour in some configurations but give poor convergence in others. The user should test the behaviour of such elements with the form of distortion that he wants to use in his analysis.

The constant stress condition is necessary for convergence since, in the limit, if an element is made infinitesimally small it will then correctly represent the stress at a point, which is constant. Both convergence conditions can be investigated using the patch test as discussed in chapter 5. For this a small group of elements are connected together into a simple geometry and this is subjected to various load cases that should reproduce either strain free rigid body motions or constant stress conditions. Although the test geometry is deliberately chosen to be of a simple external shape the internal mesh within this outline is distorted to test the response of the element to such geometrical changes. A typical patch test mesh for a membrane or a plate bending element is shown in Figure 7.13. The exterior shape is taken to be rectangular but the interior mesh consists of distorted quadrilateral elements. The shape of the mesh in this example is chosen because it is typical of the distortion that is employed for mesh density changes used for quadrilaterals. It is also arranged so that the internal nodes have a different number of elements connecting to them, that is the nodes have a varying valency. Other internal geometries can be used to test for other forms of distortion.

To test for strain-free rigid body movements, boundary movements corresponding to rigid body displacements are applied all around the perimeter of the mesh and the internal displacements checked to ensure

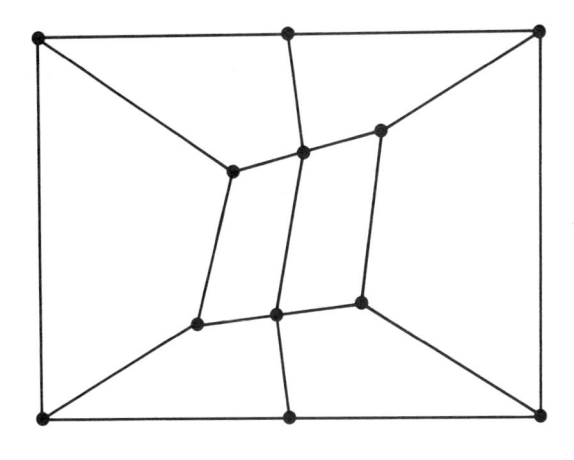

Figure 7.13

that the complete patch moves as a rigid body. The stresses within the elements should then be zero for this loading. It will be found that the calculated stresses are not exactly zero but have some very small value caused by the rounding error associated with the finite computer word length for all of the numerical operations involved. In this case there can be a difficulty in deciding if these numbers are computed zeros or are actually significant stresses. One possible test to assess their significance is to reverse the sign of about half of the applied boundary displacements so that they no longer represent a rigid body movement. The order of magnitude of the stress that arises from this can then be compared with that of the stresses which should be zero. If these differ by more than $n/2$ (where n is the computer word length) the stresses can be taken as computed zeros. For an element that can lie in a three dimensional space, even if the element itself is a line or a surface, there are six possible rigid body displacements, three translations and three rotations. All six rigid body displacements must be tested. The rotations can be taken about any axes, although those used to define the patch coordinate system is probably the easiest to use. If the element is only two dimensional then there are three rigid body movements in the plane of the element, two translations and one rotation and three normal to the plane of the element, one translation and two rotations. For axisymmetric elements the rigid body tests are rather different. General axisymmetric elements allow the loads to be specified as harmonics of a Fourier series around the circumference of the element. For the zero harmonic (axisymmetric loadings) there are two rigid body movements, translation parallel to the central axis and torsional rotation about this axis. For the first harmonic suitable combinations of the radial and the tangential displacements (and the local nodal rotations for a thin shell) can be used to give translations in two orthogonal directions perpendicular to the central axis and rotations about these directions. Together the zero and the first harmonic give all six possible rigid body movements. Any other higher harmonics in the Fourier series do not have any rigid body movements associated with them.

The patch test geometry can also be used to test the constant stress behaviour. For a three dimensional solid there are six stress components, three direct and three shear. Distributed edge loads must be applied to give a constant value of each stress component in turn throughout the patch. All six stress components must be tested independently. Membrane elements have three stress components, two direct and one shear. A plate element must be capable of sustaining constant bending stresses in two directions and a constant twisting shear stress. The general shell must be capable of sustaining both the membrane and the plate element constant stress systems.

When approaching a new type of problem that they have no previous knowledge of, then experienced users conduct a series of small parametric tests to explore combinations of structural and element behaviour related to their problem, before the full analysis is conducted. The tests are used to find the mesh density that is needed to obtain results to the required accuracy. As an example the analysis of a complicated shell structure can be investigated by studies on a simple cylinder. There will be a stress concentration at discontinuities in the shell and the die-away length associated with this together with the most suitable mesh density can be found from simple studies. Taking the typical shell radius and thickness of the actual structure to define the test cylinder then the parametric model shown in Figure 7.14 can be used for the test. Various loadings can be considered, typically a radial shear force applied around the circumference of the shell is considered in Figure 7.14.

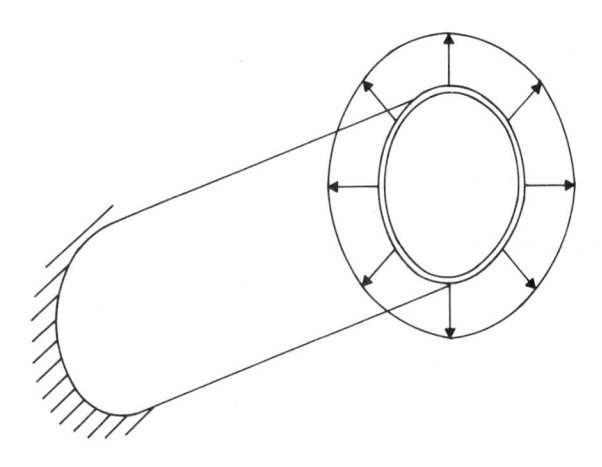

Figure 7.14

If the user has no previous knowledge of shell behaviour then he might argue that there will be peak stresses at the built in end, or at the loaded end, or be the same order at all points along the cylinders length. These alternatives can be investigated using the three meshes shown in Figure 7.15. They all have the same number of elements so that comparative assessments can be made. The radial displacements that are obtained from these three meshes are also shown in Figure 7.15. It will be seen from this

Examples of different meshes

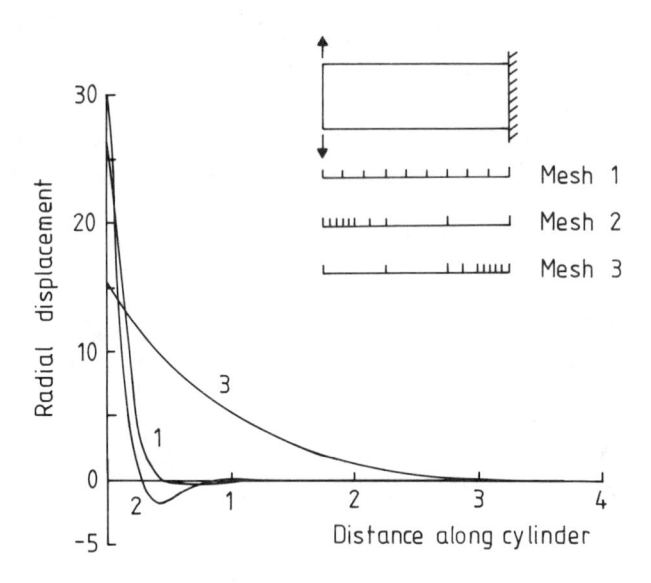

Figure 7.15

that the uniform mesh and the mesh which has more elements near the loaded end give similar results but the mesh that is denser near the built in end has a significantly different response. This indicates that the dense mesh at the built in end is not correct for this loading. Further, the mesh that is denser at the loaded end has a higher peak value and shows a higher rate of change of displacement (a sharper peak) than the uniform one. The approximations in the finite element solution are such that this is a very strong indication that the mesh that is dense near the loaded end is the best of the three for the loading considered here. The actual mesh density can be investigated further by trying various densities and different numbers of elements at the loaded end to find the most efficient mesh for this type of problem. Further parametric studies can also be conducted with different thickness to radius ratios and cylinder lengths to investigate the effect of these parameters under a range of different loadings. Such tests can be done to establish general rules for use of the element or they can be carried out for the specific geometrical forms that are involved in the problem under consideration.

The ideas developed here can be applied in a more general form. Typically, for say the analysis of a cylinder-to-cylinder intersection which is a general three dimensional problem. This can be initially investigated as a very much simpler axisymmetric cylinder-to-sphere intersection in order to establish general stress levels and mesh densities. Knowledge of the results of this simpler model gives a great detail of information for setting up the full three dimensional model. A simple cylinder can be used to find the number of plate elements that are required in the circumferential direction or the number of solid elements needed through the thickness of a more general non-axisymmetric shell. The cylinder is useful because various non finite element solutions exist for a variety of loads allowing absolute convergence tests to be carried out. Such a sequence of operations might appear, at first sight, to be time consuming and wasteful. However, in practice the careful step-by-step approach will generally lead to more reliable results in a shorter time than trying to tackle the full problem directly. This is because the analyst often has very little knowledge of what results to expect from the full analysis and therefore has no basis to assess the validity of the computations. The effort expended in checking these at least corresponds to the work required by the simplified analysis. Further, if such checking proves that the analysis is inadequate then a repeated full analysis with a refined mesh doubles the cost and the time of the analysis.

Another practical approach to analysing a complex structure is to start with a very simplified model and to gradually refine it as the design process progresses. This top down approach to the analysis means that a simple model is used when the details of the actual geometry and loadings are not well defined. As they become more precise, so the finite element model is refined. The results of each preceding analysis gives the user an insight into how to refine the model and the results that are to be expected for the next stage. Although this approach sounds rather long winded in practice it can prove to be a very efficient practical design process.

When the user is testing element behaviour to establish a suitable mesh density he should apply the types of loadings that are to be used in the final analysis. Some systems give relatively crude approximations to some of the loads that are to be applied in relation to the element interpolation. In this case a finer mesh might be required, just to apply the loads, than would be expected from the basic element convergence performance. Any tests that the user conducts should be slightly more complicated than simple rectangular geometries and constant loads since these can give a false impression of the convergence characteristics for the actual problem.

7.7 Material Properties

Within a stress analysis the principal material property is the stress strain law and it is always necessary for the user to define this. If a dynamic analysis is being conducted then the material density must also be given. Other properties may also be required depending upon the loadings to be considered. Typically the user must specify the coefficient of thermal expansion for thermal loadings and the material density for dead weight or centrifugal loads. If the model allows for material non-linearities then the associated law must be specified. For an elastic plastic analysis this requires, at the very least, a definition of the yield stress and the yield criteria that is to be used. Similarly if a visco-elastic or creep analysis is to be carried out then the properties associated with the strain-rate stress law being used must be given.

One very common mistake is in the user specifying incorrect units for the various properties, especially for the mass units. These can be checked by confirming that the product

$$\text{Young's modulus} \times \text{length} \times \text{length}$$

has the same units as

$$\text{density} \times \text{length} \times \text{length} \times \text{length} \times \text{length/time} \times \text{time}$$

and both have units of force. The first equation is stiffness times displacement and the second mass times acceleration. The units for other equations or loadings can be checked in a similar fashion.

If the material within an element is isotropic (that is the properties are not directionally dependant) and homogeneous (that is the material properties are constant throughout the material) then only one value of each material property must be given for each element. Materials can be isotropic but non-homogeneous, typically because of a varying temperature field throughout the element. In this case the user has to specify the material properties at a series of points throughout the element or, more usefully, as a table of properties against temperature (or position or any other variable) and the system automatically looks up the table and interpolates for the material property at any point. A material is non-isotropic for many reasons. Any fibrous material is naturally non-isotropic. Often some forms of construction can be modelled as a non-isotropic material rather than in full detail. Reinforced concrete in civil engineering, boiler tube sheets in mechnical engineering, stringer reinforced panels in aircraft construction and rows of bolt or rivet holes in general engineering problems are all examples of where such modelling is applicable. This is discussed in detail in chapter 12.

The majority of material properties are used directly in the form that they are measured but this is not the case with the stress-strain law. For a general three dimensional body there are six components of stress and strain, three direct and three shear. For an isotropic material the stress-strain law is defined by the two properties, Young's modulus, E, and Poisson's ratio, v. The full stress-strain law is (3.10)

$$\begin{bmatrix} \varepsilon_{xx} \\ \varepsilon_{yy} \\ \varepsilon_{zz} \\ \varepsilon_{xy} \\ \varepsilon_{yz} \\ \varepsilon_{zx} \end{bmatrix} = \frac{1}{E} \begin{bmatrix} 1 & -v & -v & 0 & 0 & 0 \\ -v & 1 & -v & 0 & 0 & 0 \\ -v & -v & 1 & 0 & 0 & 0 \\ 0 & 0 & 0 & 2(1+v) & 0 & 0 \\ 0 & 0 & 0 & 0 & 2(1+v) & 0 \\ 0 & 0 & 0 & 0 & 0 & 2(1+v) \end{bmatrix} \begin{bmatrix} \sigma_{xx} \\ \sigma_{yy} \\ \sigma_{zz} \\ \sigma_{xy} \\ \sigma_{yz} \\ \sigma_{zx} \end{bmatrix} \quad (7.1)$$

This gives the material flexibility matrix. For the displacement method the inverse of this is required and is (3.11)

$$\begin{bmatrix} \sigma_{xx} \\ \sigma_{yy} \\ \sigma_{zz} \\ \sigma_{xy} \\ \sigma_{yz} \\ \sigma_{zx} \end{bmatrix} = \frac{E}{(1+v)(1-2v)} \begin{bmatrix} 1-v & v & v & 0 & 0 & 0 \\ v & 1-v & v & 0 & 0 & 0 \\ v & v & 1-v & 0 & 0 & 0 \\ 0 & 0 & 0 & (1-2v)/2 & 0 & 0 \\ 0 & 0 & 0 & 0 & (1-2v)/2 & 0 \\ 0 & 0 & 0 & 0 & 0 & (1-2v)/2 \end{bmatrix} \begin{bmatrix} \varepsilon_{xx} \\ \varepsilon_{yy} \\ \varepsilon_{zz} \\ \varepsilon_{xy} \\ \varepsilon_{yz} \\ \varepsilon_{zx} \end{bmatrix}$$

(7.2)

which is the material stiffness matrix. There is no ambiguity in the case of a system which has three components of direct stress. However, there are two possibilities for a two dimensional stress system. In all cases the out of plane shear terms ε_{yz}, ε_{zx}, σ_{yz}, and σ_{zx} are all zero. The third direct stress, σ_{zz}, or the corresponding strain, ε_{zz}, can be assumed to be zero and this leads to two different models. The first case where the through thickness stress, σ_{zz}, is assumed to be zero is called the plane stress model. The material properties are found by deleting the third, fifth and sixth rows and columns from the material flexibility matrix of equation (7.1). The condensed matrix is then inverted to give the plane stress material stiffness matrix as

$$\mathbf{E} = \frac{E}{1-v^2} \begin{bmatrix} 1 & v & 0 \\ v & 1 & 0 \\ 0 & 0 & (1-v)/2 \end{bmatrix}$$

with the subsidiary equation

$$\varepsilon_{zz} = \frac{-v}{E}(\sigma_{xx} + \sigma_{yy})$$

defining the through thickness strain. This assumption is valid for thin sheet materials, typically an aircraft skin.

The other alternative is to assume that the through thickness strain, ε_{zz}, is zero giving the plane strain assumption. The associated material stiffness matrix is then found directly by deleting the third, fifth and sixth rows and columns from the material stiffness matrix of equation (7.2). This gives

$$\mathbf{E} = \frac{E}{(1+v)(1-2v)} \begin{bmatrix} 1-v & v & 0 \\ v & 1-v & 0 \\ 0 & 0 & (1-2v)/2 \end{bmatrix}$$

with the through thickness stresses found from the subsidiary equation

$$\sigma_{zz} = \frac{vE}{(1+v)(1-2v)}(\varepsilon_{xx} + \varepsilon_{yy}).$$

It can be important for the program to calculate this stress component where any flow failure criteria is being used since it can significantly affect the hydrostatic stress within the material. The plane strain assumption is used where a long three dimensional body is being modelled as a two dimensional slice, typically in modelling foundations in civil engineering problems.

For non-isotropic properties the stress-strain properties are directionally dependent and for the general three dimensional case 21 independent quantities, ϕ_i, must be given to define the material flexibility matrix as

$$\mathbf{F} = \begin{bmatrix} \phi_1 & & & & & \text{symmetric} \\ \phi_2 & \phi_3 & & & & \\ \phi_4 & \phi_5 & \phi_6 & & & \\ \phi_7 & \phi_8 & \phi_9 & \phi_{10} & & \\ \phi_{11} & \phi_{12} & \phi_{13} & \phi_{14} & \phi_{15} & \\ \phi_{16} & \phi_{17} & \phi_{18} & \phi_{19} & \phi_{20} & \phi_{21} \end{bmatrix} . \tag{7.3}$$

It is relatively common to have no coupling between the direct and shear stresses and strains and in this case only nine independent material properties are required so that the material flexibility matrix is

$$\mathbf{F} = \begin{bmatrix} \phi_1 & & & & & \text{symmetric} \\ \phi_2 & \phi_3 & & & & \\ \phi_4 & \phi_5 & \phi_6 & & & \\ 0 & 0 & 0 & \phi_7 & & \\ 0 & 0 & 0 & 0 & \phi_8 & \\ 0 & 0 & 0 & 0 & 0 & \phi_9 \end{bmatrix} . \tag{7.4}$$

The terms ϕ are found experimentally by applying unit stresses in each of the six components in turn with the stresses in the other components set to zero. Each loadcase gives six strains and these give the corresponding column in either equations (7.3) or (7.4). The material stiffness matrix is found by inverting the material flexibility matrix of either equations (7.3) or (7.4). Plane stress or plane strain assumptions can either be built into the experimental method used to obtain the material properties or by manipulating the (6×6) material flexibility matrix exactly as was done for the isotropic case.

The user must take care when specifying non-isotropic properties to ascertain if they are to be given in terms of the global structural coordinate directions or in the local element coordinate directions. It is also necessary to determine if the direction of the non-isotropic properties vary over the element.

7.8 The Use of Symmetry

There are many occasions when the structure to be analysed has various symmetries. These can be used to reduce the cost and to simplify the analysis. The use of symmetry is especially relevant for a dynamic analysis when a modal solution is employed since the mode shapes themselves will either be symmetric or anti-symmetric. The existence of structural symmetries increases the chance of the structure having repeated resonant frequencies but solving for symmetric and antisymmetric modes separately will generally eliminate problems associated with finding equal roots. Mode shapes can be identified directly as being either symmetric or antisymmetric rather than two arbitrary linear combinations of these. The main problem with the use of symmetry is associated with the results assessment. This must be generalised to allow for both symmetric and antisymmetric combinations and, because of this, it requires more thought and organisation on the part of the user. The problem does not exist if the loadings themselves also exhibit the same symmetries as the structure. If this is the case then the symmetries should always be exploited in the analysis.

There are various forms of symmetries that can occur in a structure, and the main forms are:
a. Mirror image symmetries, as illustrated in Figure 7.16a.
b. Axisymmetries, where the structure is obtained by rotating it about a central axis. This form of symmetry is illustrated in Figure 7.16b. It is very common since such axisymmetric shells are easy to manufacture and structurally very efficient.
c. Cyclic symmetry as illustrated in Figure 7.16c. This is reminiscent of axisymmetry but here the structure is composed of repeated sectors around the axis rather than rotating a cross-section about the axis.
d. Repetitive symmetry, as illustrated in Figure 7.16d. In this case the structure is composed of continuously repeated sections. In practice this will not occur since the structure must start and end. However, the end effects can often be ignored and repetitive symmetry used to analyse the main body of the structure.

Combinations of these forms of symmetries can also occur. The loadings on the structure can also have symmetries, in which case the solution is somewhat simplified. Alternatively it can be more general, in which case it has to be considered as the sum of a series of separate load cases where each term in the series is either symmetric or antisymmetric. The use of symmetry with a general loading relies upon the validity of the principle of superposition and this is only true for linear problems. Structural symmetries can only be used in non-linear problems if the loading exhibits the same symmetry as the geometry. Symmetry conditions are included in the analysis in two ways. Either the user defines the geometry

MESH SPECIFICATION

(a) Mirror symmetry

(b) Axial symmetry

(c) Cyclic symmetry

(d) Repetitive symmetry

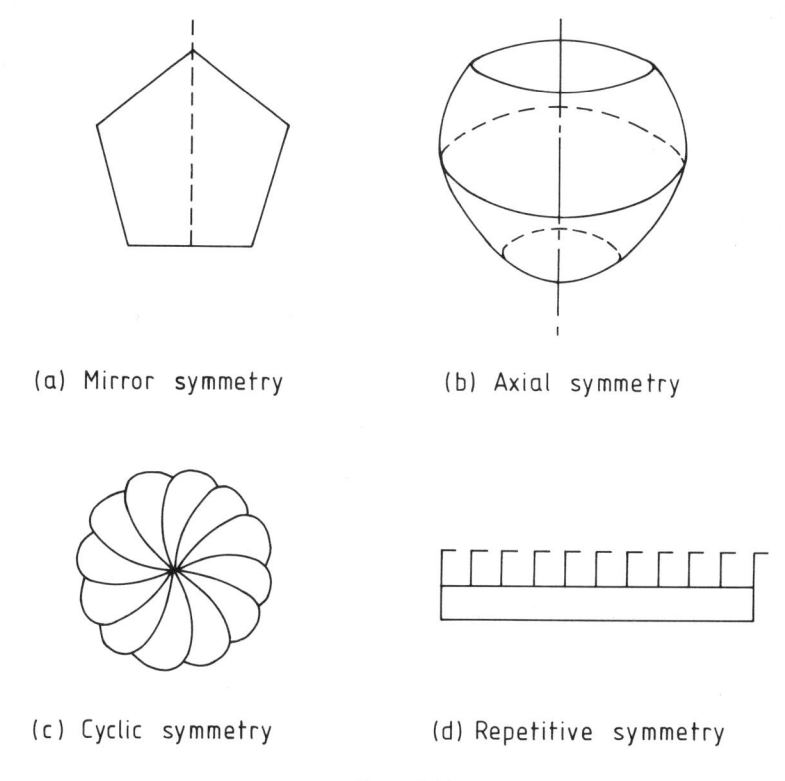

Figure 7.16

in such a way that displacement boundary conditions can be applied along the lines of symmetry, or the symmetry is included in the formulation of the element itself when the shape functions are defined. This is normally only used for axisymmetry.

A typical example of mirror symmetry is shown in Figure 7.17. It is easier to specify the relevant boundary conditions if the user defines the geometry such that the face of symmetry is normal to a global coordinate axis. In the example in Figure 7.17 the plane of symmetry is normal to the y-axis. The loading is general and has to be split into symmetric and antisymmetric components, as shown in Figure 7.18.

The section properties of any members that lie along the line of symmetry are halved because only half of the member is active in the model analysed. The other half is active in the second half of the structure which

A FINITE ELEMENT PRIMER

FULL FRAME STRUCTURE

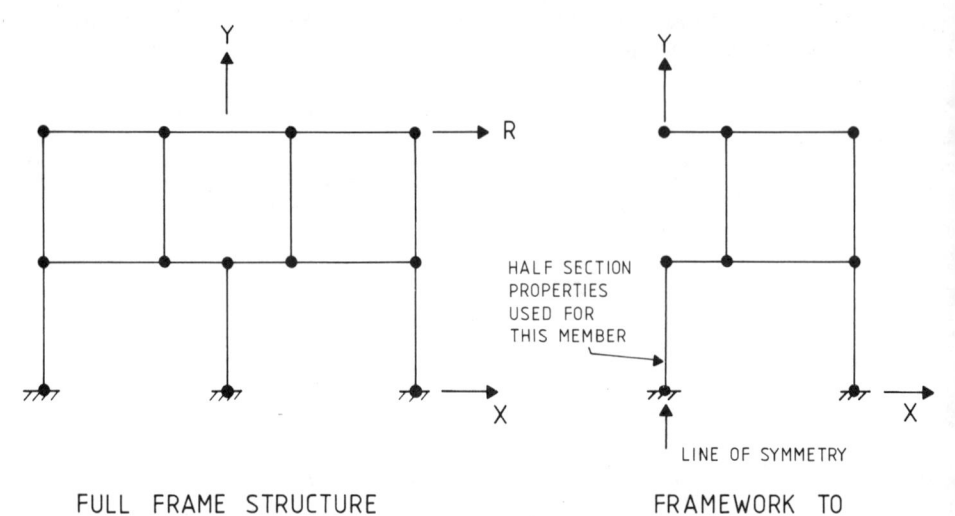

FRAMEWORK TO BE ANALYSED

HALF SECTION PROPERTIES USED FOR THIS MEMBER

LINE OF SYMMETRY

Figure 7.17

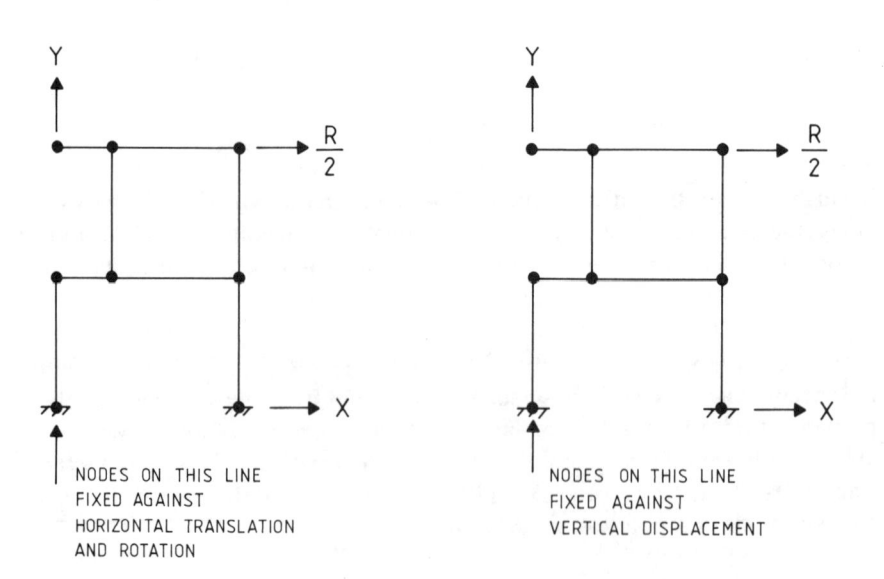

NODES ON THIS LINE FIXED AGAINST HORIZONTAL TRANSLATION AND ROTATION

NODES ON THIS LINE FIXED AGAINST VERTICAL DISPLACEMENT

Figure 7.18

118

is not being analysed. The full response of the structure is found by solving first for the symmetric load components with conditions of displacement symmetry applied along the y-axis and then solving for the antisymmetric load components with conditions of antisymmetry specified. The conditions for symmetry and antisymmetry about any of the three global axes are shown in the following table (u, v, w are displacements parallel to the x, y and z global axes respectively and uu, vv and ww are the rotations about these axes).

Plane of symmetry	Symmetry					
	u	v	w	uu	vv	ww
xy	FREE	FREE	FIX	FIX	FIX	FREE
yz	FIX	FREE	FREE	FREE	FIX	FIX
zx	FREE	FIX	FREE	FIX	FREE	FIX
	Antisymmetry					
	u	v	w	uu	vv	ww
xy	FIX	FIX	FREE	FREE	FREE	FIX
yz	FREE	FIX	FIX	FIX	FREE	FREE
zx	FIX	FREE	FIX	FREE	FIX	FREE

The structure can have symmetry about more than one axis. If there is symmetry about two axes then the possible combinations of boundary conditions are symmetric/symmetric (SS), symmetric/antisymmetric (SA), antisymmetric/symmetric (AS) and antisymmetric/antisymmetric (AA). For the general three dimensional case with eight symmetries then there are eight combinations of boundary conditions and eight sets of results. Before these can be solved the general loading must be decomposed into the loads required for the various boundary conditions. All of the forces parallel to the x-axis in the positive x, y, z quadrant can be denoted by $R1(+++)$, in the negative x and positive y, z quadrant by $R1(-++)$ and so on for each quadrant. The forces parallel to the x-axis corresponding to the symmetric/symmetric/symmetric boundary conditions are $R1(SSS)$, for the antisymmetric/symmetric/symmetric boundary conditions $R1(ASS)$ and so on for each set of boundary conditions. These two sets of forces are related by

$$Q1 = A1 * R1$$

119

where

$$Q1 = \begin{bmatrix} R1(SSS) \\ R1(ASS) \\ R1(SAS) \\ R1(AAS) \\ R1(SSA) \\ R1(ASA) \\ R1(SAA) \\ R1(AAA) \end{bmatrix} \quad R1 = \begin{bmatrix} R1(+++) \\ R1(-++) \\ R1(+-+) \\ R1(--+) \\ R1(++-) \\ R1(-+-) \\ R1(+--) \\ R1(---) \end{bmatrix}$$

and

$$A1 = \begin{bmatrix} I & -I & I & -I & I & -I & I & -I \\ I & I & I & I & I & I & I & I \\ I & -I & -I & I & I & -I & -I & I \\ I & I & -I & -I & I & I & -I & -I \\ I & -I & I & -I & -I & I & -I & I \\ I & I & I & I & -I & -I & -I & -I \\ I & -I & -I & I & -I & I & I & -I \\ I & I & -I & -I & -I & -I & I & I \end{bmatrix}$$

where I is the unit matrix equal in size to the number of nodes in the quadrant that is being analysed. The above relationship also holds for moments about the global x-axis for frame and shell problems. The forces parallel to the global y-axis (and moments about this axis) can be transformed in a similar form as

$$Q2 = A2 * R2$$

where

$$A2 = \begin{bmatrix} I & I & -I & -I & I & I & -I & -I \\ I & -I & -I & I & I & -I & -I & I \\ I & I & I & I & I & I & I & I \\ I & -I & I & -I & I & -I & I & -I \\ I & I & -I & -I & -I & -I & I & I \\ I & -I & -I & I & -I & I & I & -I \\ I & I & I & I & -I & -I & -I & -I \\ I & -I & I & -I & -I & I & -I & I \end{bmatrix}$$

and the forces and moments in the z direction are

$$Q3 = A3 * R3$$

where

$$A3 = \begin{bmatrix} I & I & I & I & -I & -I & -I & -I \\ I & -I & I & -I & -I & I & -I & I \\ I & I & -I & -I & -I & -I & I & I \\ I & -I & -I & I & -I & I & I & -I \\ I & I & I & I & I & I & I & I \\ I & -I & I & -I & I & -I & I & -I \\ I & I & -I & -I & I & I & -I & -I \\ I & -I & -I & I & I & -I & -I & I \end{bmatrix}.$$

These relationships obviously simplify considerably for structures with fewer symmetries and the combinations for these can be derived from the above general relationship. Although the use of symmetry requires the solution of a series of boundary conditions and loading cases this is generally considerably cheaper from a computational point of view than the solution of the very much larger full problem. The penalty paid for the increase in efficiency is the extra care and effort that the user must expend in handling all of the combinations of boundary conditions and loadcases. If this is handled automatically by the analysis program then the user should always take advantage of any structural symmetries. The handling problem is considerably simpler for the calculation of natural frequencies and mode shapes since this does not involve any loadings and symmetry should always be used in such cases because of the cost of the eigenvalue extraction.

Structures that are axisymmetric are usually best solved by means of axisymmetric elements. These are specifically formulated by defining suitable shape functions over a cross section which is then rotated about an axis to form a toroidal element. By this means it is possible to solve problems that are three dimensional using only a two dimensional section thereby giving a considerable saving in the cost of the analysis. In the general case the loading on the structure is not axisymmetric and has to be expanded as a Fourier series in terms of the angle θ around the circumference of the shell. The load at any angle is

$$R(\theta) = \sum_{n=0}^{m} A_n \cos n\theta + \sum_{n=1}^{m} B_n \sin n\theta.$$

A FINITE ELEMENT PRIMER

It can be shown (Chapter 6) that for linear problems all of the harmonics in this series can be solved as a separate problem. The geometry is the same for all harmonics but the stiffness matrix is a function of the harmonic number, n, and has to be reformulated each time. If the loading is axisymmetric then only the harmonic number zero term in the Fourier series need be solved. For those problems where relatively few terms are required in the Fourier series the process is very efficient when compared to other methods of modelling the structure.

The displacements associated with a node of an axisymmetric element can be a translation in the radial direction, a translation in the axial direction, a rotation about an axis tangential to the circumference of the shell and a translation in the same direction. (The other two rotations are automatically defined by the Fourier series). If the first three displacements varies as $\cos(n\theta)$ then the tangential displacement varies as $\sin(n\theta)$. Similarly if the first three displacements vary as $\sin(n\theta)$ then the last varies as $\cos(n\theta)$. The stresses also vary around the circumference. The three direct stresses and the shear stress in the plane of the element vary as $\cos(n\theta)$ [$\sin(n\theta)$] and the two remaining shear stresses vary as $\sin(n\theta)$ [$\cos(n\theta)$]. The maximum values of the stress components are usually printed so that the user must take care to use the correct combinations when obtaining equivalent and principal stresses at any angle around the circumference.

The beam bending behaviour of an axisymmetric shell can be recovered by using harmonic number one. If the radial and tangential displacements of a node have the same value (or possibly equal and opposite depending upon the way that the element has been implemented) then the circumference at that node stays circular for harmonic one. Similarly making radial and tangential forces equal (or opposite) corresponds to applying a pure shear force. An axial force which varies as $\cos\theta$ has a resultant which corresponds to an overall bending moment.

Cyclic symmetry is rather like axisymmetry in that the displacements around the circumference are expanded as a Fourier series. However, since the structure is not axisymmetric it cannot be solved in terms of the behaviour of a single cross-section. Instead the geometry of a typical sector is modelled. The process of cyclic symmetry then relates the displacements on one edge of the sector to those on the opposite edge. The load is expressed in terms of a Fourier series and the various harmonics solved. The displacements and stresses within the structure can be found by solving for each harmonic and recombining these according to the basic Fourier series. This requires a separate combination for each sector of the structure. The use of cyclic symmetry can be very beneficial for specific types of structure but it does demand a significant degree of understanding on behalf of the user and should be used with caution. It is rather easier to use, and has significant advantages, for finding resonant frequencies of cyclically symmetric structures. The user should check if the angle of the sector is defined in a clockwise or anti-clockwise direction and ensure that his specification of the problem follows the convention defined by the program. Such checking is especially necessary where the mesh is being generated by a pre-processor which does not know about cyclic symmetry.

Repetitive symmetry is closely related to cyclic symmetry. Again the displacements are expanded in terms of a Fourier series and this used to relate edge displacements of a typical segment. This method should be used with caution and only by users that have a good appreciation of how the structure is going to behave. Very few systems have cyclic symmetry or repetitive symmetry facilities.

8. Assembly and Solution

8.1 Introduction

Once the mesh for a finite element analysis has been constructed then the analysis itself can be performed. This requires much less input from the user than the mesh preparation and at this stage it is the computer that is doing the work. A flow chart for a typical finite element stress analysis is shown in Figure 8.1. Not all, if any, of the systems currently in use follow this sequence and some have considerably more steps. However, Figure 8.1. does illustrate one very important aspect of the method in that the solution proceeds in a series of well defined steps. Any one step uses data generated by previous steps and is itself generating data that can be used subsequently. This means that the finite element process is naturally modular and most general pupose systems exploit such modularity. It is one of the fundamental reasons as to why large scale general purpose finite element programs can be developed. Each module is written and checked almost as a separate entity, with only a minimal of interaction between modules. The second reason for the development of general purpose systems lies in the fact that many of the operations used in the FE method are identical in form and the same computer code can be used repeatedly. Any 'bugs' in the program are much more likely to be found if the same section of code is used in different parts of the program and the fewer the lines within a program the more reliable the code is likely to be.

The modular nature of the FE method also allows an analysis to be stopped and restarted at various stages, usually between modules. (Note that some systems refer to modules whilst others speak of steps or phases. The terms are not synonymous since their actual meaning depends upon the implementation used within a particular system. However, they all loosely mean the same thing and are taken to be the same in the context used here.) The possibility of using restarts is enhanced since most systems have some form of internal analysis database associated with the solution process and this can be saved and used as a restart file. It is easiest for the

ASSEMBLY AND SOLUTION

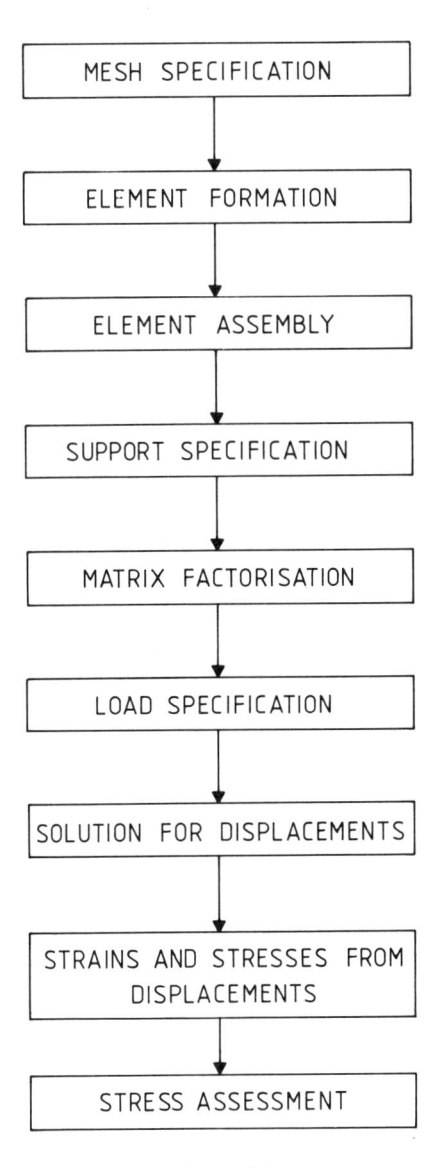

```
┌─────────────────────┐
│  MESH SPECIFICATION │
└──────────┬──────────┘
           ▼
┌─────────────────────┐
│  ELEMENT FORMATION  │
└──────────┬──────────┘
           ▼
┌─────────────────────┐
│  ELEMENT ASSEMBLY   │
└──────────┬──────────┘
           ▼
┌─────────────────────┐
│ SUPPORT SPECIFICATION│
└──────────┬──────────┘
           ▼
┌─────────────────────┐
│ MATRIX FACTORISATION│
└──────────┬──────────┘
           ▼
┌─────────────────────┐
│  LOAD SPECIFICATION │
└──────────┬──────────┘
           ▼
┌─────────────────────┐
│SOLUTION FOR DISPLACEMENTS│
└──────────┬──────────┘
           ▼
┌─────────────────────┐
│ STRAINS AND STRESSES FROM │
│     DISPLACEMENTS   │
└──────────┬──────────┘
           ▼
┌─────────────────────┐
│  STRESS ASSESSMENT  │
└─────────────────────┘
```

Figure 8.1

user if the database can be saved as a single physical file, but some systems require the management of a collection of files for restart purposes. This is very much less convenient than a single database and the user has to implement a sophisticated automatic file handling facility if restarts are to be used extensively over a range of different analyses that are all being conducted concurrently. If this is done then mistakes in handling the files are minimised. Such considerations are not apparent when a system is used on a one off basis in a development phase but they can be dominant when it is put into production use since they can give rise to so many quality assurance problems.

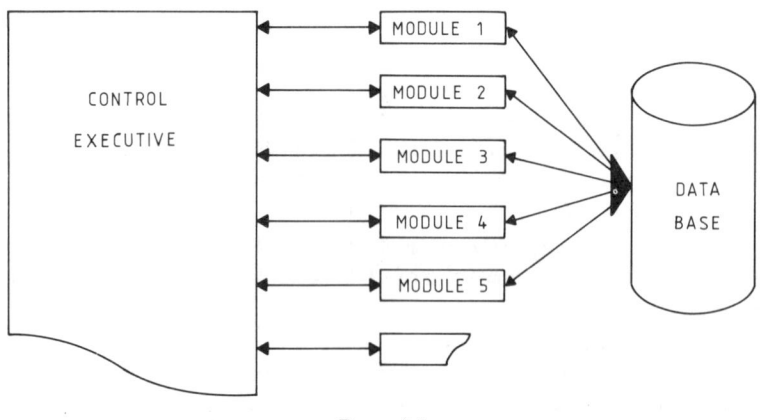

Figure 8.2

The architecture of a typical system that exploits the modular nature of the FE method and using a single analysis database is shown in Figure 8.2. Here the program execution is controlled by a central executive that calls up the various modules that are required to carry out a stress analysis. Each module reads data from, and writes data to, the analysis database as required. Such an architecture minimises the number of intermodule links within the program which should ensure that it has a long useful life because its development potential is then open ended.

8.2 Checking Procedures

There are three phases for checking a finite element analysis:
a. Before the analysis – This is carried out at the mesh generation and the matrix formation stages.

b. During the analysis – This checks for consistency of data and for numerical conditioning.
c. After the analysis – To investigate the results and to try to confirm that the answers are correct.

The first checking phase, before the analysis, is largely done at mesh generation. However, the user should take care if a separate pre-processor program has been used to generate the mesh. Pre-analysis checks should always be carried out within the FE system itself since there can be errors in the interface step between the pre-processor and the analysis. This usually arises because of differences in interpretation between the two programs. It is important that the FE program conducts the checks so that the user is informed of the analysis that is actually being conducted and not just what the pre-processor has been instructed to do.

Checks during the analysis will find two types of error. They can be fatal, in which case the execution is terminated, or they can be warnings which indicate inconsistencies. It is important that such warnings are not completely suppressed since they are often the only indication of an error that the system gives before producing meaningless results. They are often imbedded within the main output but it is helpful if they are also copied to an error log file so that they are summarised for the user. If this is available then they can be suppressed in the main body of the output. Warning messages can often be dismissed since they will be triggered by some pre-set tolerance that might not be applicable to the current analysis. However, it is the users responsibility to confirm that warnings can be ignored.

Post analysis checks are usually carried out by the user outside of the analysis. They are concerned with verifying that a given set of results are sensible and often involve hand calculations or other simple sums to confirm that the order of magnitude of the results are correct. The availability of plotting facilities is also useful here to confirm that the general trends within the solution are correct.

8.3 Element Generation and Assembly Problems

Various difficulties in the use of the FE method can arise at the initial phase of the analysis. These problems usually fall into one of three categories:
a. Excessive element distortion.

A FINITE ELEMENT PRIMER

b. Incorrect element connections.
c. Incorrect mixing of element types.

FE systems will contain tests and diagnostics for at least one of three possible errors. It is usually impossible to flag them as absolute errors since their degree of importance depends upon the problem being solved. A common phrase used by the developers of FE programs is that they assume 'an intelligent user'. In reality what this phrase means is that the user is expected to be very conversant with the fundamentals of the FE method and is able to understand and interpret any warning diagnostics that the system produces. (Worse still are the systems that actually assume that an intelligent user is a full blown expert with vast experience in that they produce very few or no warning diagnostics.) A good system will have various levels of diagnostics ranging from elementary ones that assume the user is a total novice up to those appropriate for an expert. The various levels can be suppressed on the output but they should always be printed on a separate error log file. This then enables any reviewer of the analysis to look at the error log first when vetting the analysis but it still preserves a tidy output file free of messages that are not significant. Given the choice it is better to have a system print a series of diagnostic messages that the user has to explain rather than no messages at all when there actually are errors.

8.4 Excessive Element Distortion

The formulation of all finite elements will ultimately be based upon the assumption of some basic geometrical shape for the element. Typically for an isoparametric quadrilateral the basic shape is a square, since this is used to map onto the real distorted geometry. For a triangle the basic shape is usually an isosceles triangle. As the element is distorted from this basic shape to the shape in the real space then possible errors occur in the associated mapping. The more distorted the element the less well it will model the behaviour of the structure. The basic forms of element distortion are defined in section 4 of chapter 7. There are various methods in use for measuring these distortions, some of which are general and some are particular to certain elements. The significance of any distortion will always depend upon the actual element since higher order ones are generally more tolerant to distortions.

There are three possible forms of distortion:
a. The volume of infinitesimal elements are not mapped uniformly throughout the finite element when transforming from the basis space to the real space.

b. The element has a high aspect ratio so that the length of one or more sides is very much greater than the length of the shortest side.
c. The element is skewed or tapered.

For high order elements that can curve in space then these definitions apply at the infinitesimal level but they still have the same geometrical meaning. In this case the distortion for the element can be taken as the worst value anywhere within the element. This will usually be at a node point but tests are often only conducted at the Gauss integration points since this is where the element functions are formed.

All of these forms of distortion can be quantified by means of the element Jacobian (see equation (5.18)) at any point. This relates the derivatives of functions in the basis space to derivatives of the same functions in the real space. For a three dimensional solid the Jacobian has the form:

$$\mathbf{J} = \begin{bmatrix} J_{11} & J_{12} & J_{13} \\ J_{21} & J_{22} & J_{23} \\ J_{31} & J_{32} & J_{33} \end{bmatrix}.$$

The value of the determinant of this matrix is

$$\det(\mathbf{J}) = J_{11}J_{22}J_{33} + J_{12}J_{23}J_{31} + J_{13}J_{21}J_{31} - J_{21}J_{22}J_{13} - J_{32}J_{23}J_{11}$$

$$- J_{23}J_{21}J_{12}$$

and this gives a measure of the volume distortion. For no volume distortion then $\det(\mathbf{J})$ will be constant at all points within the element.

Each row in the Jacobian can be interpreted as the direction of the local mapping vector,

$$\mathbf{v}_1 = [J_{11} \quad J_{12} \quad J_{13}]$$

is the direction vector of a line of constant ξ_1 in the real (x, y, z) space. Similarly

$$\mathbf{v}_2 = [J_{21} \quad J_{22} \quad J_{23}]$$

$$\mathbf{v}_3 = [J_{31} \quad J_{32} \quad J_{33}]$$

are the direction vectors of constant ξ_2 and ξ_3 respectively. The local length, l_i, of the ith vector is then

$$l_i^2 = (J_{i1}^2 + J_{i2}^2 + J_{i3}^2)$$

and the aspect ratio distortion can be found by comparing the ratio of the three vector lengths l_1, l_2 and l_3. For zero-aspect ratio distortion these ratios are always unity,

$$AR = \max\left(\frac{l_1}{l_2} \frac{l_2}{l_3} \frac{l_3}{l_1} \frac{l_2}{l_1} \frac{l_3}{l_2} \frac{l_1}{l_3}\right).$$

The angular (skew and taper) distortion can be found by evaluating the angles between the vectors. The cosine of the angle between the i and j vectors is given by

$$AD = \max\left(\frac{\mathbf{v}_1 \mathbf{v}_2^t}{l_1 l_2} \frac{\mathbf{v}_2 \mathbf{v}_3^t}{l_2 l_3} \frac{\mathbf{v}_3 \mathbf{v}_1^t}{l_3 l_1}\right)$$

With quadrilateral and cubic elements then this term is zero for no angular distortion and for any type of element the closer it is to unity then the more distorted is the element.

For axisymmetric elements the volume distortion should include the radius of the point under consideration but if this is done then an excessive degree of distortion will always be flagged if the element touches the axis of symmetry. The terms in the stiffness matrix do become large in this case so the fact that a distortion error is flagged is meaningful. However, in practice the solution can be arranged to be well conditioned in this case and often the radius is not included in the volume distortion measure for axisymmetric elements.

Methods of testing for element distortions have some problems associated with them, mainly concerned with defining a meaningful value for the tolerance used in defining what is a significant distortion. Some such tolerance must be used to trigger an error message but if it is too coarse then quite acceptable distortions are flagged as possible errors, but if it is too fine then serious distortions can be missed. The value of valid tolerances will vary with element types, basically the higher the order of the element interpolations then the more distorted it can be. Systematic tests must be conducted for all element types within a system in order to establish valid tolerances. Alternatively the tolerances can be established for the most commonly used elements and assumed to be valid for all element types. Users should conduct their own tests to establish distortion tolerances for the types of elements, geometries and load cases that they most commonly use.

For these reasons most distortion tests can only be used as a guide to possible errors since the consequences of the distortion also depends upon the loadings. Typically, if the stresses within an element are substantially constant then a much higher distortion can be accepted than for those cases where the stresses are varying rapidly. Element distortions should be minimised in those regions of the mesh where the stresses are of most interest, or where stress concentrations occur. Away from such regions much higher distortions can be tolerated. This can be especially useful so far as element aspect ratios are concerned since these can be made relatively high in the less important parts of the mesh. (This is not the case for a dynamic analysis where wave propagation effects are of interest and in this case aspect ratios of one must be used.)

8.5 Incorrect Element Connections

Within the displacement method care is taken in the element formulation to ensure that continuity is preserved everywhere. However, the user is still at liberty to violate continuity with his choice of the mesh. Whilst he must be free to do this where necessary the program should still provide some feedback that continuity has been violated. This can be important where automatic mesh generation or mesh pre-processing facilities have been used, since if the nodes are closer than some tolerance, then they will be connected. Some systems contain checks that flag any nodes which are close together but not actually connected, say those which are within 10 times the connection tolerance. The user can then check to see if these close nodes should in fact be interconnected. This has been found to be very useful in large three dimensional solid models that have curved surfaces and variations in mesh density simultaneously in all three directions. The interpolations used in modelling the curvatures can lead to unconnected nodes and faces in such cases, and the resulting mesh is so complicated that the user must have feedback from the program that indicates possible regions of error.

Compatibility can also be violated by elements being connected together incorrectly, as discussed in section 7.3. A typical problem that occurs here is the case when a small number of nodes in a mesh have been left unconnected since the generating program placed them outside the 'capture' tolerance (see section 7.2). Such errors can be detected by assigning a status to a node the first time that it is encountered. This status can classify the node as being either corner, edge, face or internal. When the same node is encountered on other elements their status can be checked again and if it is not the same as the original definition then it can be flagged as a possible error. This check will find errors where mid side nodes are connected to corner nodes but it will also find the majority of cases where elements of different order of interpolation are connected or where nodes on a single element have been collapsed onto each other. It is possible to assign a code for different forms of element interpolation functions which then allows the program to check that only compatible elements are connected together. This is rarely done in practice.

8.6 Incorrect Mixing of Element Types

Most systems allow the user complete freedom in the choice of elements used in the mesh, but not all elements are compatible with each other. There are often good modelling reasons for using incompatible elements, but on other occasions elements are mixed by error or by a lack of experience on the part of the user. The degree to which this checking can be conducted depends upon how the system that is being used has been written. At one extreme (probably the most common situation) no checking at all is done, whilst at the other the system flags all element types and only allows compatible elements to be mixed. Intermediate stages between these two extremes can be used. For example, if two elements to be joined at a node have a differing number of freedoms per node then there is a possibility of a lack of compatibility and this can be signalled. This then indicates that the user must confirm that the correct freedoms have been matched.

Some systems have elements of different spatial dimension, with separate two and three dimensional forms of an element. In all systems some elements are axisymmetric. Incompatibilities will occur if elements with a different spatial dimension are mixed and a common fault here is to mesh two dimensional membrane elements with axisymmetric ones. This almost invariably leads to models that have no physical meaning and delivers fictitious results. If such a model is attempted then the user must first determine if the axisymmetric element is defined as a complete ring or as a one radian slice or as some other sized slice. The membrane element thickness and other properties are then chosen to smear their effects. This type of modelling is probably better done using only axisymmetric elements with modified material properties.

8.7 Graphical Mesh Checks

It is almost impossible to verify a mesh without access to a large range of graphical facilities. Two dimensional and axisymmetric meshes can be checked reasonably efficiently with only hard copy plotters. This is not so for three dimensional meshes and these require extensive use of interactive graphical facilities. It is very important that these graphical checks are carried out before the analysis and not as a part of it. Almost every mesh of any real structure contains errors and a great deal of man time and computer time is saved if any mesh has been carefully verified before it is used. The minimum graphics requirements are:

a. A high resolution plotting device, typically with a 1024×1024 pixel resolution. Any lower resolution makes it impossible to present the fine detail of a mesh.
b. Hard copy facility, preferably a reasonable quality pen plotter or a high resolution screen dump.
c. Element plotting with facilities for either numbering nodes and elements or for switching off these numbers to give a clear plot.
d. Selective plotting of elements, either in a defined window, or by selected groups of elements.
e. Plotting of the mesh from any viewpoint. True perspective plotting can be useful here but parallel perspective is sufficient.
f. Selective scaling in any coordinate direction, to allow the details of high aspect ratio structures to be investigated.
g. Element shrinking or outline drawing.

For interactive plotting it is desirable to have as many facilities available as possible since they will all be useful at some stage. Features that are often useful include:

a. Menu driven operation with control of operations by either cursor control or from the keyboard as appropriate.
b. Options to plot curved element edges as either straight line segments for rapid plots or in their true curved form for distortion checking and final report results presentation. Elements should always be plotted using the element shape functions themselves rather than any other form of curve fitting since this will show the true geometry that has been analysed rather than just drawing a 'pretty picture' representation of the mesh.

A FINITE ELEMENT PRIMER

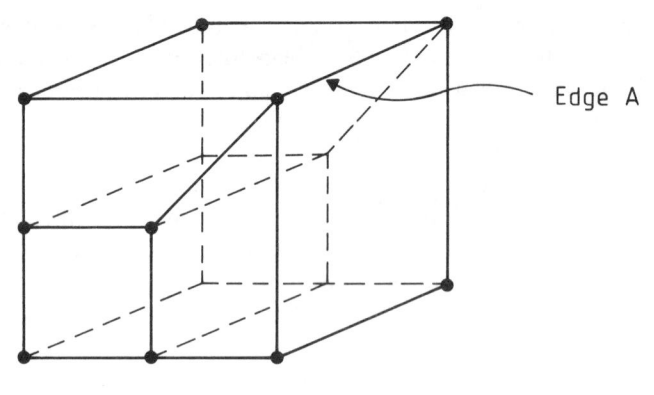

Figure 8.3

c. Various forms of outline plotting. A simple algorithm can be used for outline definition, where a line is designated as an edge if it only occurs on one element. Although this gives a rapid method for detecting most outlines it is not always correct. For example, the mesh of brick elements shown in Figure 8.3 will not have the edge marked A drawn if the simple edge detection algorithm is used, since it occurs on two elements. A better procedure for detecting edges is to first determine the surface faces as being those element faces that are not common to any other element. An edge can then be defined as being either edges that only occur on one element, or as a common line between two faces, where the normals of the faces differ by more than a specified amount. This information is also required for hidden line removal algorithms. Such data, if it is available, can also be useful for checking boundary condition and loading data. Usually displacements will only be fixed on the surface of the structure and some load types, typically pressures and tractions will also be restricted to external faces.

d. Hidden line removal, to give a plot only of those surfaces which can be seen from a specified viewpoint. This usually only produces a pretty plot at the mesh generation stage and is not of much use for verification purposes. A true outline plot, as discussed in the previous paragraph is almost invariably much more useful for verification. However, hidden line removal, or at least the identification of visible element faces is necessary for contour plotting on the surface of three dimensional meshes. A simple form of hidden line removal can be obtained on raster-scan graphics terminal by drawing the mesh in order from the rear of the view to the front and using a blank fill within each face. This will leave only the visible edges in the view. Unfortunately this is of no help for contour plotting since the visible faces are not defined explicitly and the program does not know where to draw the relevant contours.

e. The indication of close, but unconnected nodes. This aids the user in detecting disjointed meshes and internal 'cracks'.
f. Facilities to obtain node and element numbers in small regions of the mesh. The extent of these regions can be specified by the graphics cursor. This feature is useful for identifying node and element numbers for subsequent material and load specifications.
g. Plotting true outlines for beams, pipeworks and plates. These are defined in terms of centre lines and often only the centre lines can be plotted and this can lead to ambiguity in the interpretation of the plot.
h. Graphical representation of beam principal axes and rigid offsets for beams and plates. The definition of the principal axis direction is always difficult and error prone for anything but the simplest sections. Any checking facilities for this will make the use of beam models considerably simpler and more reliable.

This list is not complete and almost any facility is useful at some time. It is not possible to verify a mesh using graphics alone and the program checks discussed in section 8.5 and the graphical presentation are used together. Typically a single unconnected node will not be noticeable in a graphics display since it will only show as a single dot. In practice it is the three dimensional meshes, especially solid element meshes that require most aids for verification. These meshes are often so complicated that they must be viewed in parts and the user needs some guide from the program as to which portions of the mesh should be investigated graphically.

Some typical analysis plots taken from actual jobs are shown in Figure 8.4 to 8.8. Figure 8.4 shows a bracket and gives plots both with and without hidden line removal (not from the same viewpoint). Even though the geometry is relatively simple it will be seen that the plot without hidden line removal is rather complicated and not very informative. The hidden line removal plot is easier to read and some rather elongated wedge elements can be seen. Obviously this does not allow the interior of the mesh to be checked for consistency. Figure 8.5 shows a double cylinder model, again both with and without hidden line removal. The plot without hidden line removal also illustrates the shrunken element facility giving an 'exploded' view of the model. These plots are both from the same viewpoint and, as such, they can be misleading. Figure 8.6 plots the same structure with a very slight change in the eye point and it shows

A FINITE ELEMENT PRIMER

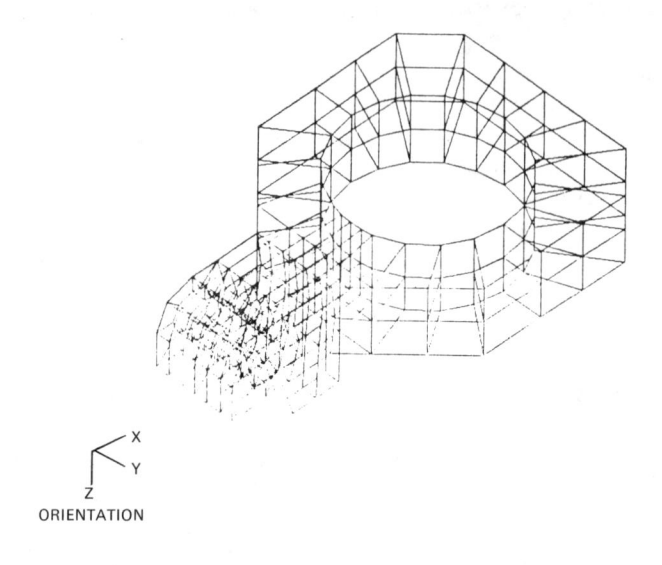

Figure 8.4

ASSEMBLY AND SOLUTION

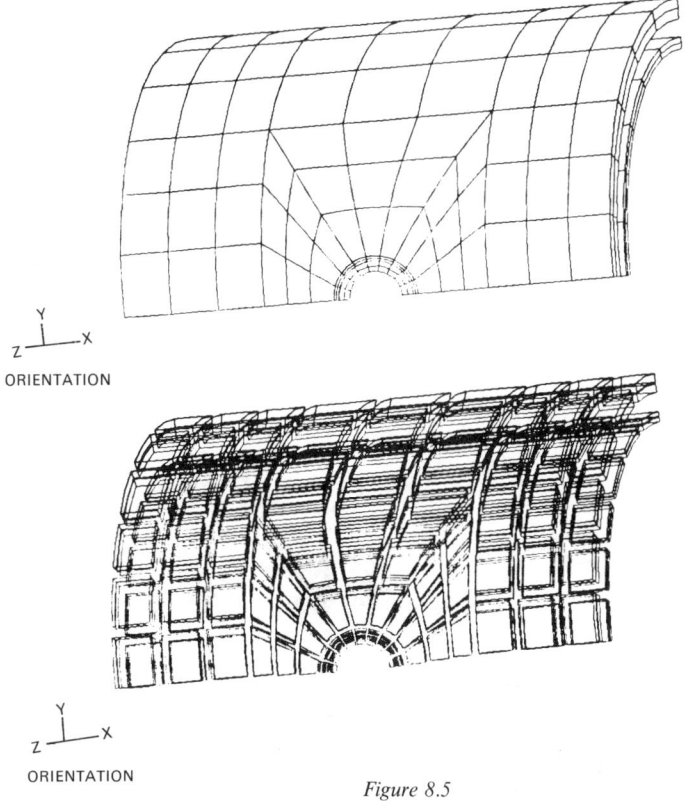

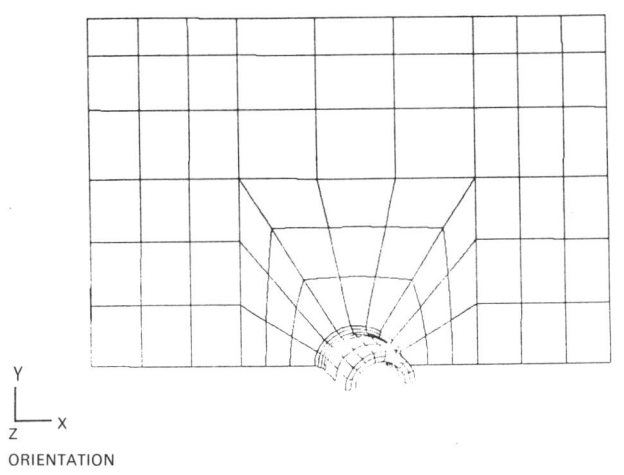

Figure 8.5

Figure 8.6

A FINITE ELEMENT PRIMER

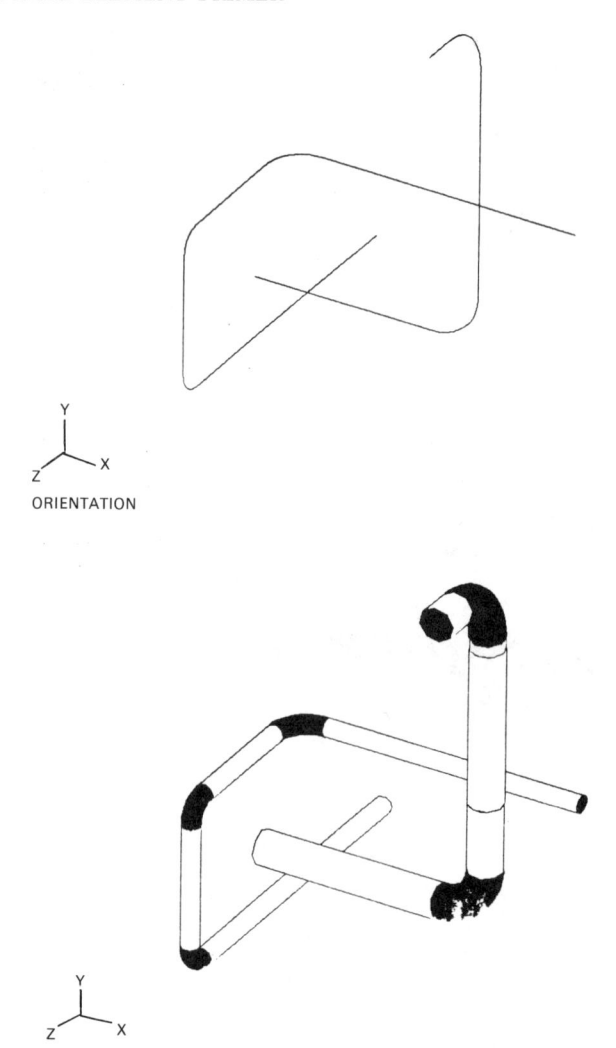

Figure 8.7

that there is an inclined nozzle penetrating the two cylinders. Figure 8.7 gives a plot of a part of a pipework system. The first plot in the figure shows it as a centre line drawing and it is not clear from this if the two sections are separate pipes or are interconnected. The second plot shows the same pipework with actual diameters for the pipe, from which it is clear that the two sections do not intersect. This plot was obtained from a screen dump of a colour display where bends and pipe cross-sections are

138

Figure 8.8

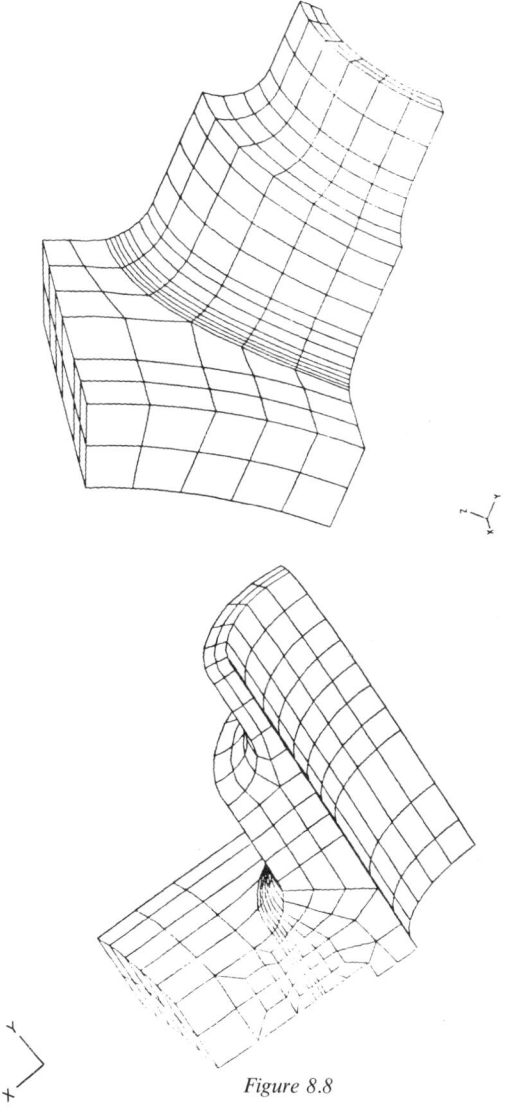

plotted in a different colour from the straight sections, hence the dark areas on the plot. Figure 8.8 shows plots of a nozzle/cylinder intersection. The first picture shows a relatively standard mesh but the second shows that a high degree of mesh refinement has been used in the region of the connection. The mesh density is reduced by 'paving' the elements away from the dense mesh region. The paving occurs in all directions and at the same time the geometry is curving in all three dimensions making the mesh very difficult to generate. It required the use of all of the automatic checking facilities described in this section before all of the errors in the mesh were eradicated. This included both graphical checks and a large number of checks at the mesh generation stage within the finite element program. In practice the debugging of this model took longer than the time required in its initial generation.

8.8 Ill-conditioning in the Finite Element Method

Any arithmetic operation in a digital computer is prone to rounding error since the computer can only work to a fixed number of significant figures. The rounding error usually arises when the difference is taken of two similar numbers. The significant figures then cancel and the result is determined by the less significant and less accurate digits. Within the computer a number is stored in a given number of binary bits, called the word length of the computer. Some machines with an inherently short word length allow the programmer to combine two or more words together so that numbers can be stored in double or multiple precision making more significant figures available for computations. There are in fact two separate numbers within this word length. A floating point number is stored in the form of a mantissa and an exponent. For example the number 12,340 ($=.1234E5$) has a mantissa of .1234 and an exponent of 5. A fixed number of bits of the computer word is reserved for the mantissa and the rest is used for the exponent so that a different precision can be achieved between two computers even when they have the same word length. This occurs if a different balance is used for the storage of the exponent and the mantissa. For the most part compilers increase the size of both the mantissa and the exponent as the word length is increased but there are some machines that use the same number of bits for the exponent irrespective of the actual word length and this can cause problems. Users should always determine the word length within the FE system that they are using (and how both the exponent and the mantissa are stored) since it will control the basic accuracy of the system.

In general all calculations should be carried out with at least a 60 bit precision. If a shorter word length is used then it is found that many problems can be solved accurately but some have errors arising from roundoff due to the lack of a sufficient number of significant figures. It is difficult to predict problems that will trigger this form of error. Some three dimensional problems with many thousand degrees of freedom can be solved accurately with a short word length whilst some shell problems, with only about two hundred equations, can fail because of roundoff error. It is also necessary to carry out all of the calculation with at least 60 bit precision since errors due to roundoff can occur at any stage of the computation. Typically there is a temptation to find the stresses from the displacements using a short word length since this reduces the cost of the calculation. If a thermal stress analysis is being conducted then the elastic strain is found by subtracting the thermal strains from the total strains and this involves taking the difference of two very similar numbers with the inherent possibility of ill-conditioning.

8.9 Solution Diagnostics

Any static stress analysis problem requires the solution of a series of simultaneous equations. For large problems there can be many thousands of equations and, over the past 15 years various efficient solution techniques have been developed for handling such large sets of equations. The finite element method leads to a heavily banded symmetric set of equations. Although it would seem at first sight that such a sparse set are best solved by some form of iterative method, in practice it has been found that the rate of convergence of such methods is very unpredictable for the equations that are generated by the finite element method. This makes it an unreliable technique for general purpose programs and it is rarely used, although some recent developments might revise this conclusion for large scale non-linear problems. In practice some form of direct solution of the equations is used. There are two widely used solution methods Gauss elimination and Cholesky factorisation.

Both of these methods start by taking the stiffness matrix, **K**, and decomposing or factorising this into the product of two triangular matrices. For Gaussian elimination this takes the form

$$\mathbf{K} = \mathbf{LU}$$

where **L** is a lower triangular matrix and **U** is an upper triangular matrix. In practice this is solved in two steps, firstly forward elimination

$$\mathbf{Ur} = \mathbf{L}^{-1}\mathbf{R} = \mathbf{y}$$

followed by a backward substitution step

$$\mathbf{r} = \mathbf{U}^{-1}\mathbf{y}.$$

The matrix **L** is never formed explicitly and it can be inferred from **U** because the stiffness matrix is symmetric. The backward substitution step is very easy to carry out because **U** is an upper triangular matrix.

The Cholesky factorisation method is closely related except that it uses the fact that the stiffness matrix is symmetric and positive definite. Under these conditions the matrix can be factored in the form

$$\mathbf{K} = \mathbf{L}\mathbf{L}^t.$$

This time the displacements are found by both a forward and a backward substitution step, but both of these are quick to perform because the matrices involved are triangular.

Gauss elimination is used in conjunction with a frontal solution process, where the element assembly and equation triangularisation (forward elimination) are combined so that a full set of assembled equations are not formed. Instead the equations are assembled and as soon as it is possible to factorise an equation then this is done, effectively forming the factorised matrix directly. With the Cholesky factorisation method the elements are assembled to form the structure's stiffness matrix as the first step which is then factorised forming the triangular matrix as the second step. Both techniques have advantages and disadvantages with respect to each other but these are relatively minor. Provided that both methods have been programmed efficiently then they cost substantially the same to run and both have the same degree of accuracy. Both techniques have proved to be significantly superior to any other solution method.

In factorising the stiffness matrix or the forward elimination step a very large number of numerical operations are performed. These mainly consist of multiply and add operations (inner products), with a series of divisions associated with the diagonal terms of the matrix. The multiplications usually present no difficulty unless the exponent of the computer word is very short. It is the additions that are subject to rounding error. If the two terms to be added differ in sign and are almost equal this gives the difference between two similar numbers and significant round off errors can be introduced. If the error occurs with the diagonal term then subsequent divisions using this has the effect of magnifying and propagating the error. It would seem from this that an unlikely series of events are required to cause such errors but unfortunately the structure of the stiffness equations are such that they are very likely to occur. A single diagonal term in the stiffness matrix gives the stiffness of the corresponding degree of freedom with all other of the degrees of freedom fixed. This will always be a large number relative to the lowest stiffness of the complete structure. However, the final deformed shape of the structure is much more likely to be defined by this overall stiffness, rather than the stiffness at any point and the final diagonal terms of the factorised matrix will represent this low stiffness. The factorisation process causes the size of the leading diagonal terms to get smaller and this arises through a sequence of cancellations of similar numbers which can lead to rounding errors. The solution process will always be susceptible to such rounding errors.

The same conclusion can be arrived at from other considerations. The deformed shape, $\mathbf{r}$, can be shown to satisfy the equation

$$\tfrac{1}{2}\mathbf{r}^t\mathbf{K}\mathbf{r} - \mathbf{r}^t\mathbf{R} = \text{a minimum}$$

where the left hand side is the difference between the internal strain energy and the external work. Formal minimization of this function leads to the standard stiffness equations. Minimising the strain energy involves taking differences of terms in the stiffness matrix until the minimum is found. As this is approached then the operations get progressively more ill-conditioned.

From these arguments it will be seen that a measure of possible roundoff error can be obtained by considering the highest possible stiffness that the structure can have compared to the lowest possible. The ratio of these two numbers is called the condition number of the stiffness matrix. If the order of the condition number approaches the length of the number of significant figures that the mantissa of the computer word can represent, then the factorisation is likely to be incorrect. It does not follow that the solution is necessarily wrong since, so far, the actual loadings have not been considered. For example, if the structure has no supports then the lowest possible stiffness it can have is zero corresponding to strain free rigid body movements. However, if the applied loads are self equilibrating then the rigid body motion movement will not occur and it is quite likely that a perfectly good solution is obtained. On the other hand, if the loading is not self equilibrating then the solution will be meaningless.

The condition number of a stiffness matrix can be found exactly by taking the ratio of its highest to lowest eigenvalue. Unfortunately this is not of very much use in practice since the effort required to find these eigenvalues is considerably greater than that required to solve the equations. Instead many programs calculate some approximation to the condition number. There are a variety of such approximate measures. It can be shown that the sum of the leading diagonal terms of the stiffness matrix (the trace of the matrix) is equal to the sum of its eigenvalues. Hence an upper bound on the highest eigenvalue can be found from trace of the stiffness matrix. The approximate lowest eigenvalue can be considered as the smallest diagonal term in the factorised stiffness matrix. The ratio of these two numbers will then give the approximate condition number. If the order of magnitude of this is less than half the number of significant digits that the mantissa of the computer word can represent then there will be no significant rounding error. If the order of the condition number lies between a half and three quarters of the number of significant figures then there is a definite possibility of an error in the solution and if it is greater than this then there is a very strong likelihood of error.

It is reasonably common for programs to either print the condition number as a standard part of the factorisation process or to print a warning message if it exceeds a pre-defined value. In such cases the user should check the value of the condition number and if it is becoming large with respect to the computer word length then the results should be treated with caution and be investigated thoroughly by the user to verify that they are sensible. The user should also check for other simple modelling errors, typically that the structure is correctly supported, that the material properties have been correctly specified and that any modelling idealisations are in fact reasonable.

Another common form of checking at the solution stage is to calculate the diagonal decay. This is the ratio of each term on the leading diagonal of the stiffness matrix after triangularisation with the value before. If the order of magnitude of the diagonal decay is tending to zero for any term then this indicates a possible loss of accuracy since it gives a measure of the number of significant figures lost in the calculation. If more than half the precision has been lost then the results should be thoroughly checked for consistency.

The user can identify structures that are likely to be susceptible to rounding error as being those that have both very high components of stiffness and very low components within the same model. The simplest example of this is the beam element that has both bending and end load terms. The bending stiffness will usually be orders of magnitude less than the axial and it is possible to obtain inaccurate solutions, even in what appears to be relatively trivial frame or pipework problems. This is even

ASSEMBLY AND SOLUTION

more true for a plate or a shell analysis. If a shell is represented using thick or solid elements then the chance of numerical error is further compounded by the very high through thickness stiffness that such structures will have. If a structure does have regions of disparate stiffness then the user should try to improve the accuracy of the answer either by choosing an element where the high stiffness has been eliminated in the modelling (say using a plate element rather than a solid brick) or by applying generalised constraints (multi point constraints) to constrain out the areas of high stiffness. This type of conditioning problem can also occur where the user specifies a very fine mesh, either all over the structure or just in one or two regions. The lowest stiffness associated with the structure is defined by its overall geometry and is independent of the mesh, where as the highest stiffness is essentially defined by the distance between the two closest nodes. As the mesh is made finer so the conditioning of the solution process gets worse.

The condition number or the diagonal decay only give indications of the possibility of errors in the solution. Whether there are actually any errors then depends upon the load case. Most programs provide a check at the end of the solution process to verify that the applied forces are balanced by the product of the stiffness times the displacement, that is after solution the norm $\|\mathbf{Kr}-\mathbf{R}\|$ is shown to be very much less than the norm, $\|\mathbf{R}\|$, of the applied forces. The fact that this is not satisfied does not necessarily mean that there are significant errors in the displacements or the stresses. It can be shown that a very small error in the displacement can lead to a much larger error in the effective force found by multiplying this displacement by the stiffness. The multiplying factor can be as high as the highest eigenvalue of the stiffness matrix. The result of this test must be treated with some caution by the user and used as an indicator of the level of checking that is required. If the program diagnoses both a condition number error and a residual force balance error then it is very likely that the solution is unacceptable. If only either a condition number or a residual force balance error is given then it is likely that the solution is correct. In all cases it is up to the user to exercise his judgement to decide whether or not to accept a solution.

8.10 Other Program Checks and Diagnostics

General purpose finite element systems usually have a variety of other diagnostic tests built into them that are carried out as a part of the analysis. These vary from system to system and some are more useful than

A FINITE ELEMENT PRIMER

others. At the load formation stage it can be convenient to have the program print out the total resultant loads and moments with respect to the origin of the global axes for the load case. This gives the user an easy check, especially in a qualitative sense. The user can quickly verify that the signs of these resultants are correct and that the loads have been applied in the correct coordinate directions. This can be very useful for cases where the loads are applied in a local direction, typically pressures applied normal to a face of the structure. The sign of these will be associated with the local directions and the method in which the local directions are defined vary from system to system. Some programs do not have convenient methods of defining local directions and a printout of the resultants in global directions allow independent checks to be made.

Another set of simple checks that the program can produce are volume or mass calculations for individual elements and the complete structure. This has usually been computed for other reasons such as material ordering and costing and gives the user a quick check on the basic validity of the model. Having the volume or mass available for individual or small groups of elements, allows specific areas of inconsistency to be detected. Typically if there is a wrong specification of plate thickness then this will not usually be shown on any graphical tests but it will give an error in the total volume or mass. It can also be useful to have the moments of inertia of the structure computed and printed since these also give quantities that can be checked independently.

It is convenient for the user if the program flags the highest displacement and stress components since this will ensure that the important aspects of the results are not missed. Such a requirement might seem to be redundant at first sight since the user can find this for himself from the results. However, for some analyses, especially with three dimensional solid elements, where there are thousands of nodes and many thousands of stress values there is just too much data for the user to assimilate without help from the program. The user should know where he expects to find the regions of high stress and largest displacement before the analysis is started. If he does not know this, he cannot devise a suitable mesh. If the peak values occur in other parts of the structure then either there is an error in the mesh, in the input data or the user does not understand the behaviour of the structure and needs to do more exploratory work before a reliable mesh can be generated.

8.11 Results Presentation

The main aim of the analysis is to verify that the structure can do its job.

This usually means that either some displacement, or stress or strain limits must be satisfied. Often these are not directly related to the output from the finite element program, typically the structure must be assessed against some code of practice and the code has different stress categories for the same stress depending upon the loading case that generated the stress. In such cases it can be convenient for the user to have a separate postprocessor to carry out these code checks. Most general purpose programs do not include code checks since the code of practice will vary from industry to industry and sometimes from company to company. If the results are to be assessed against a code of practice then this should be considered when the mesh is initially defined since a suitable choice of elements and mesh can considerably simplify the consequent code checks. If the code is based upon thin shell theory and gives stress categories in terms of end load and bending moment then, unless there are good reasons otherwise, shell elements should be used for the analysis. If brick elements are used then great difficulties can occur in the stress assessment since the program will deliver much more detail in the results than the code is designed to deliver. It is also quite likely that the increased freedom that is available for generating a solid mesh means that there is probably not a suitable face of elements for obtaining overall end load and bending moments. The forms of stress output that are commonly available are discussed in detail in the next chapter.

8.12 Improving Solution Efficiency and Accuracy

A two dimensional and axisymmetric finite element solution is relatively cheap compared to a solid or general shell analysis and the user can make serious underestimates of the cost and the time required for the analysis if extrapolations are made from two dimensional to three dimensional problems. As a working rule of thumb it has been found that the cost of carrying an analysis of a structure using a 3-D mesh is in the order of 20 to 50 times that of the same structure used with a 2-D model. The man time and associated costs of preparing the mesh and investigating the results are also increased by about the same amount. It will be appreciated from this that it is important to strive for efficiency when solving a 3-D problem. Within the analysis program the modules that give rise to the cost vary according to how the program was written. The two steps that are the most expensive are those associated with the element formation and the equation factorisation. So far as the user is concerned he has relatively little control over the element formation stage. The majority of the cost is incurred in the sum total of all of the element integrations, especially the formation of the product **B′EB**. (See section 5.4.) For a 20-node solid element **B** is a (6 × 60) matrix and the product gives a (60 × 60) matrix. Even allowing for the symmetry of the equations this will still contain some 1830 terms. The product has to be formed at each integration point and, even with reduced integration, there are still eight points per element. It will be seen from this that just the formation of a single element requires a considerable amount of number crunching. If full integration is used then the product has to be evaluated at 27 points, that is three times as many evaluations are required as for reduced integration. This is one of the reasons for the popularity of reduced integration (the other being that it tends to give improved accuracy) since, even if twice as many elements are required with reduced integration, there is still a reduction of integration time. In practice the same mesh can usually be used for reduced integration as for full. If the user has any control over the number of integration points then the lowest recommended number should be used. Care must be taken not to reduce the number too far otherwise the elements contain zero stiffness modes that do not have zero strains and consequently the solution will become meaningless. It is also important from a cost point of view that the minimum number of elements consistent with the required accuracy is used. However, it must be recognized that this usually implies a graduated mesh and it becomes more difficult to set up and verify a variable density three dimensional mesh, so that this method of reducing the solution cost usually increases the mesh generation man time. It is important here to have carried out some initial studies on approximate two dimensional or axisymmetric models in order that the user has a very good idea as to what constitutes a reasonable mesh. The user should also take care over the choice of element. General shell elements are more expensive than equivalent 2-D elements but, where they are applicable, they are considerably cheaper to use than solid elements.

The user has rather more control of the analysis cost during the matrix factorisation stage. How this control is exercised depends upon the solution algorithm that is employed. If Cholesky factorisation is used then the numbering of the structure's node points is important and most general purpose mesh generation and solution programs contain routines that automatically optimise the node numbering according to some objective function. There are three meaningful objectives that can be used:
a. To minimize the bandwidth of the equations. This makes the difference between the highest and lowest node numbers as near equal and as low as possible for all elements.
b. To minimize the profile of the equations. This makes the sum total of

ASSEMBLY AND SOLUTION

the off-diagonal terms, not including the leading zeros in each row, as small as possible.

c. To minimize the number of multiplications in the Cholesky factorisation. The aim here is to reduce the number of expensive operations in the factorisation process.

The most efficient implementations of the Cholesky factorisation use a variable bandwidth (a skyline) storage scheme. Bandwidth minimization is not very relevant in such cases. Generally the minimization of the number of multiplications in the Cholesky factorisation is the best objective function to use but it is not available in many systems in which case the profile reduction can be used. This can also be the best objective function if it results in a very significant reduction in the number of terms in the stiffness matrix that have to be stored since the number of disc transfers to backing store will be reduced.

If automatic node renumbering is used it is more convenient for the user if this renumbering is completely internal and transparent to the user. Some systems provide a renumbering that literally renumbers the node both internally and externally. This means that the user gets use to a certain node numbering sequence when the mesh is being generated but the results are presented using a totally different set of node numbers. This is confusing and error prone the first time that it is used and the problems are compounded if, as a result of the analysis, the structure has to be modified slightly and this results in a third completely different numbering scheme.

If a frontal solution is used then it is not relevant to renumber the nodes. Instead it is more important to renumber the element sequence to minimize the number of terms in core. This is usually, but not necessarily, closely related to bandwidth reduction. There will always be fewer elements than nodes in a mesh and it might be expected that the element numbering optimization will be more efficient than node renumbering.

Any renumbering algorithm will take time to carry out. If the mesh has been generated automatically then the resulting node sequence will be far from optimum and a renumbering step will pay for itself many times over. This becomes less true as the element complexity is increased. Generally all renumbering algorithms work well for simple two noded line elements but as the number of nodes per element increases then so does the cross coupling in the stiffness matrix and consequently the optimization process becomes more difficult. There are some renumbering algorithms that only

A FINITE ELEMENT PRIMER

work with simple linear elements and the user should confirm that any given algorithm does in fact work with higher order elements if these are being used.

Renumbering works on a purely topological basis with no consideration as to the conditioning of the set of equations that are being solved. Usually this makes little difference but care should be taken where the conditioning of the equations is marginal. It has been shown, even on small problems that it is better to solve a set of equations corresponding to the stiffest part of the structure and moving to the most flexible part so that the equations in this region is solved last. If the equation numbers have been optimized on a topological basis then the solution sequence is not under the control of the user. Such considerations are not normally relevant but if conditioning errors do arise and these are due to high relative stiffness values in the structure then it might be worthwhile changing the renumbering objective function or switching it off altogether and ensuring that the numbering goes from high to low structural stiffnesses.

The conditioning of the equations can also be improved in some cases by scaling the stiffness matrix in the form

$$\mathbf{K} = \mathbf{DAD}$$

where $\mathbf{D}$ is a diagonal matrix

$$\mathbf{D} = [K_{11}^{\frac{1}{2}}, K_{22}^{\frac{1}{2}}, \ldots, K_{nn}^{\frac{1}{2}}]$$

and a typical term in $\mathbf{A}$ is

$$A_{ij} = \frac{K_{ij}}{D_{ii} D_{jj}}.$$

This will make the leading diagonal of $\mathbf{A}$ have only unit values. The solution is then

$$\mathbf{r} = \mathbf{D}^{-1} \mathbf{A}^{-1} \mathbf{D}^{-1} \mathbf{R}.$$

Obviously, since $\mathbf{D}$ is diagonal and its inverse is trivial. The scaled matrix $\mathbf{A}$ is not inverted but factorised in the usual way. The conditioning of $\mathbf{A}$ will be better than $\mathbf{K}$ especially for structures that have large variations in stiffness. These variations have been filtered into $\mathbf{D}$ and the inverse of diagonal matrix has no numerical stability problems. The main disadvantage of this technique is the cost of forming the scaled matrix **A** from the original stiffness matrix. The process only has advantages for ill-conditioned equations. It will not make a truly singular matrix non-singular.

8.13 Use of Program Restarts

The majority of large systems contain restart or checkpoint facilities that allow the user to halt the analysis at various stages of its execution. This can be a very useful facility, especially when carrying out large 3-D analyses. Instead of trying to carry out the analysis in a single run, and assuming no mistakes in the data, it is conducted in a series of steps. The results from each step and, especially, any diagnostic messages that are generated, are checked before proceeding to the next step. If any error is found it is corrected immediately and the stage where it occurred is rerun. This allows the user to correct errors as they occur and minimizes the running costs. It makes sense to identify the computer intensive stage of the analysis and to arrange restarts before and after these. The most expensive stages of a static analysis are the matrix formation and factorisation steps.

Restarts also have various other incidental advantages. The restart files can be copied and saved at each stage giving a measure of security to the data. The copies of the restart files can also be used for parallel analyses, say where the structure is symmetric and the loads are both symmetric and antisymmetric. The formation of the stiffness matrix will be the same for both cases and a restart file generated after the assembly step can be used as the root for the two types of boundary conditions. The restart procedure comes into its own where the structure is subjected to a variety of load cases, not all of which are fully defined at the beginning of the analysis. New load cases can be applied using a restart. The restart also allows the user to schedule the job through the machine. The computer intensive steps can be run in batch mode, probably at night or over the weekend, and the shorter steps, such as solving for a new load case or finding the stresses, can be solved inter-actively during the day. There can also be difficulties associated with the use of restarts. Some systems allow them but requires the user to manage a collection of computer files all associated with the same restart. This makes the process very cumbersome and unfriendly. More modern systems use a single database file for the restart and this is very much easier to manage. On those systems with

A FINITE ELEMENT PRIMER

only a single database file the restarts are user friendly and can be used much more freely. Another obvious problem with the use of restarts is that the restart file has to be saved. These can be very large for 3-D jobs and take up a significant amount of disc space. The degree to which this is a problem then depends upon the volume of disc space that is available. The files can be saved on magnetic tapes but this is not a very practical solution because of all of the problems associated with handling tapes.

8.14 Substructuring

Substructuring is a facility where the actual structure is analysed as a series of smaller structures. These are used to generate reduced size matrices for each component, where the reduced matrices are only defined in terms of the connection freedoms. The reduced matrices are assembled and the displacements found along the common boundaries. The boundary displacements are then fed back to the individual components to find the detailed component response. Obviously, from this description, the substructuring method is not straight forward and requires the user to save details of individual components, essentially as restart files. The degree of user difficulty involved in carrying out a substructured analysis then depends critically upon the implementation within the FE system.

Substructuring is very useful where the complete structure is made up of a series of identical smaller components. The stiffness matrix for one component can then be used repeatedly to build up the complete structure. The technique can also be very useful where a detailed stress recovery of only a small part of the structure is required, say at the tip of a crack. The results have only to be recovered for one sub component. An ideal case for substructuring is the method of virtual crack extensions for finding stress intensity factors. Here repeated runs have to be made with small changes to the crack length and a very fine mesh is required in the region of the crack tip. This region can then be constructed as a separate component and it is only this small area that has to be reformed as the crack is moved. This can give savings of a factor between 20 and 50 compared to repeating the full analysis each time. In other cases the extra work involved on the part of the user to employ substructuring negates its effectiveness. For a static analysis it can be shown that substructuring leads to an exact solution and can be considered as another form of equation solving. The condition number of the solution accumulates through the various levels of the substructures and each level is more ill-conditioned than the previous levels. In dynamics substructuring is only approximate and should be used with great caution. Substructuring is useful in a non-linear analysis where the non-linearities are confined to small, well defined regions of the complete structure, say the contact problem at a structural joint. It is not an applicable technique for general non-linear response calculations.

One form of substructuring that is available in some systems is called the super element method. In this the full structure is analysed as a series of subcomponents and these are then assembled as though they were large elements, hence the name of super elements. The technique differs in implementation from standard substructuring in that the complete process is carried out as one analysis rather than forming and saving the individual component matrices. This makes it very much easier for the user since he does not have to worry about saving the individual files for each component but it does restrict the usefulness of the method since it is not easy to modify individual components. It is used mainly for very large problems that are too big to be analysed by a single factorisation. Any form of substructuring also has organisational advantages in that each substructure can be meshed and checked independently of the rest of the geometry.

9. Results Processing

9.1 Introduction

Having carried out the finite element analysis it is necessary for the analyst to assess the results that have been obtained. It is an unfortunate fact that the user must set up and debug his mesh before the analysis can be conducted, but he can accept the results as they are presented with no equivalent debugging. There is a great temptation to assume that the results are correct after all of the effort that has been expended to obtain them. Unfortunately with the approximate nature of the FE method there is a chance that the mesh used is not sufficiently accurate to deliver the required results or that, even with the most conscientious of checking, there are still input data errors. This means that the user has the final responsibility for checking out the results that have been obtained. It is good user practice to consider a set of results to be wrong until it has been positively proved that they are at least consistent and that the expected orders of magnitude of displacements and stresses have been obtained. All of this has to be done in addition to processing the results that are actually required. In practice the engineer will probably spend more time in results checking and processing than was spent in mesh generation. This fact is often overlooked with insufficient time being allowed in the job specification for detailed results investigation and this is where the danger of accepting the unverified computer output arises. It is always a good idea to have the results assessed by a second person who is not directly involved in the project so that a critically objective view is obtained regarding their validity. Such a process is often included, at least nominally, as a part of the Quality Assurance procedures used by many companies but, to be useful, they must be on-going as the job progresses and not just occur as a final check at the end. It must also be recognized that the person who does the reviewing is required to have a detailed knowledge of the purpose of the analysis, of the fundamentals of the finite element method and the ability to find inconsistencies almost instinctively. Such people are obviously highly qualified and their talents are likely to be required at many other stages of the job. If this is the case then they are not likely to be as objective as is required, in which case a review by an outside organisation should be considered for important work.

The basic result of the FE analysis is the deflection of the structure at the node points. Other results are then derived from the displacements. The deflections are differentiated to give the strains and from these the stresses can be found. They can also be used in conjunction with the structural stiffness to obtain the reaction forces at the support points. It is also possible to use the displacements and the element stiffnesses to obtain the internal element forces (stress resultants) rather than the stresses. There are many variations that can be used for presenting these results and it is up to the user to choose suitable combinations that allow him to check the validity of the analysis and to obtain the results required for the assessment of the structural integrity.

9.2 Displacement Results

The displacements are usually presented at the structure's node points in the direction of the global axes used to define the model. The results are used to obtain the effective flexibility at any point on the structure and are important in their own right where a maximum displacement is the design criteria. Usually the displacement results are accurate and, even with a coarse mesh, reasonable displacement values are predicted. As mentioned in section 4.2 an approximate solution based upon the displacement formulation will lead to 'over stiff' results in which the strain energy is underestimated. It is therefore expected that if the structure is loaded mechanically by, say, point loads or surface pressures then the peak displacement of the structure will be under estimated when fully compatible displacement elements are used. Under the same circumstances elements based upon the force method will over estimate the displacements. If non-conforming or hybrid elements are used then the displacement results do not generally converge monotonically as the mesh is refined and it is not possible to say if any given peak displacement is an over or an under estimate. The use of reduced integration has the same characteristics as the hybrid formulation. If the loadings are applied displacements or strains, such as support movements or thermal loads, then these comments are reversed, a conforming displacement element will overestimate peak displacements and so on.

In theory it is possible to bound the displacements by carrying out two analyses, one based upon the displacement formulation and one based upon the force method. This is rarely, if ever, done since the solutions will still not bound the stresses and these are usually of more interest to the analyst than the displacements.

It is possible to use the deformed shape, especially in the regions of highest displacements, to obtain comparative estimates of the efficiency of two different meshes. Generally where the rate of change of displacements is high then the mesh that gives the highest rate will be the better one. Obviously this is closely related to the strain values but since the displacement curve is generally smooth and continuous across elements it is often easier to look at displacement plots rather than stress or strain variations.

Displacement results can be presented either as a table of nodal values or they can be plotted. There are two basic forms of graphical representation, either a plot of the deformed geometry, usually overlaid on the undeformed geometry, or as a line plot of displacement value normal and/or tangential to a line of nodes. The first method of presentation gives an overall view of the structural behaviour but is often not useful for anything more than a general appreciation of the manner in which the structure responds and as an immediate diagnosis of gross modelling errors: typically wrongly defined supports. The second method, the graph plot, gives less detail and so is not as good for an overall view but is generally more useful in a quantitative sense. In all cases the displacement plots should be obtained using the element shape functions rather than any other form of interpolation. If the shape functions are employed then the user is presented with the exact information as the model sees it. Other forms of interpolation can sometimes smooth out or exaggerate irregularities in the displacements. If the elements allow bending, so that there are rotational degrees of freedom, then the shape function will naturally reflect these in the plots whereas other forms of interpolation often will not. When plotting deformed shapes the displacements are magnified to make them visible. In some cases the scaling factor is such that parts of the structure are 'turned inside out' and the user must recognise that the plotted deformations are greatly exaggerated.

9.3 Stress Results

One of the main reasons for conducting a finite element stress analysis is to find the distribution of stress throughout the structure. These stresses are then assessed in some way by the user to establish whether the structure can carry out the task that it is designed for and that it can survive for its required life. These requirements mean that there are usually a variety of different load cases that have to be considered and the stresses produced by each do not necessarily have the same significance or require the same form of assessment. Similarly, the stresses at various points throughout the structure do not necessarily have the same significance. For various load cases the difference in significance tends to be between normal operating loads and fault conditions. The fault conditions themselves are often sub-divided into those faults that the structure should survive and still continue to operate and those that the structure can suffer such that it fails in a safe manner but need not be in an operational condition afterwards. The classification of the fault loads is based upon the probability of their occurrence. For a loading that causes the structure to fail, proving that it will still be safe, requires a full non-linear analysis to be conducted. In other cases this is often simplified by assessing the linear behaviour of the structure against the proof stress (possibly including an assessment of displacements) for the equivalent of a fault load for which the structure should still work and against ultimate stress for the equivalent of the fail safe case.

Many industries have standard codes of practice that are used for assessing stresses. Even where there are not established codes then each industry, company or even design office will have an established stress assessment procedure. The different forms of manufacture and the different uses to which structures are put means that it is impossible to define any universal assessment method. In practice there is a very wide range of possible forms of assessment and for this reason very few programs contain detailed stress assessment procedures. It is then convenient for the FE program to contain facilities for saving the results in such a form that they can be picked up by a post processing program for assessment purposes. Some programs have such post processors for certain codes of practice. It is usually not possible to completely automate code assessments since, almost invariably, at some stage the code has to be interpreted to fit a given situation.

All of the possible load combinations means that any finite element program should provide a wide range of alternative methods of presenting the stresses. Different assessment methods can require different forms of stress. The program should also be versatile in allowing the user to combine the results of various load cases.

9.4 Element Stresses

There is some difficulty for the analyst right at the start of the stress assessment process since the element stresses can be presented in various ways. Some element types present stresses in global coordinate directions and some present them in local element coordinates. As a general rule elements that involve bending almost invariably give stresses in local element coordinates since the definition of the bending component of stress must be local to the element. Also elements that use a lower dimension for the interpolating function than for the geometry tend to be in local coordinates. For example a membrane plate element has a two dimensional interpolating function but it can be used as a surface in space to analyse three dimensional structures and the stress results must be in the local element coordinates. Most solid elements tend to present the stresses parallel to the global axes. The user must check within the program documentation to see how the stress for a given element is presented and is quite possible that if two element types are being used in an analysis then one can give global stresses and the other local.

The user's difficulty is further compounded by the position at which the element stresses are given. More often than not the maximum stress occurs at the surface of the material. There must be nodes of the finite element mesh at the surface so that the most useful place for the stresses to be computed is at the element nodes. Unfortunately, since the nodes are at extreme points of the element and since the finite element method only converges in a mean square sense the stress errors tend to be greatest at the element nodes. It has been shown in previous chapters that the stresses may be more accurate at the Gauss points because these will always be in the interior of the element. The stresses then have to be extrapolated from the Gauss points to the nodes (element surface). Extrapolation almost invariably introduces errors and, in practice, the accuracy of nodal stresses which are computed directly or by Gauss point extrapolation tends to be similar although the values produced by each method are not the same. The choice of where to recover the stresses then depends upon how the user is to use them. The one exception to this is in the analysis of non-linear problems. If the non-linearity is related to stresses (and most are) then the stresses will probably only ever be calculated at the Gauss integration points since it is prohibitively expensive to track the stress at other points and these are found by extrapolation. In the linear case FE programs should provide the user with the option of computing stresses directly at node points or by Gauss point extrapolation. Some programs also allow the user to determine stresses at other points, typically either at the centroid of the element or at any general point within the element.

9.5 Stress Averaging

A draw back to the displacement form of the finite element method is that equilibrium is only satisfied in the mean or over the element. This means that along an edge which is common to two elements the stresses are different across the edge, where they should be continuous. In particular, stresses are discontinuous across nodes no matter how they are calculated. Most programs contain facilities for averaging these nodal stresses but such facilities must be used with care. If the element stresses are in local coordinates then stresses must not be averaged when the direction of the local coordinates change. This can occur when either the geometry of the structure changes or, less obviously, where the local element topology direction changes. Care must also be taken when averaging stresses across different element types since the two elements might not have the same number of stress components or they might not be acting in the same direction. Averaging should not be carried out across elements with dissimilar material types or where there is an abrupt change in the structural geometry or loading. If the stresses are averaged in such cases the user should recognize that he is effectively smearing the effect of the discontinuity across the adjacent elements. Any system that only allows averaged stresses to be printed should be treated with some caution since the stress difference across elements can provide the user with valuable insight into the accuracy of the mesh that has been used for the analysis. In the real structure, where both equilibrium and compatibility must be satisfied, the stresses will be continuous. The discontinuity that arises in the FE solution can then be used as a measure of the accuracy of the mesh. In order to establish a measure of the stress difference across elements it can be compared with the absolute maximum stress value that occurs anywhere in the model. If the stress difference is less than 10% of the maximum stress then the model can be assumed to be satisfactory. If the stress difference is greater than 50% of the maximum value then it is likely that the mesh needs refining, at least in the regions where the difference is of this order. The higher the order of the element interpolation the smaller the stress discontinuity should be. In practice it would appear that stress difference is related to the maximum stress at any point in the structure rather than to the local stress value.

If stresses are averaged at nodes where the stress should be continuous then the average value will almost always be closer to the true value than individual nodal estimates, especially in the regions of highest stress. This means that even if nodal values are taken without averaging then the percentage error, relative to the maximum stress, will be half the percentage stress difference across nodes. These comments are based upon practical observations rather than any strong theoretical basis and, as such, should be treated as guidelines rather than absolutes. One obvious consequence of this is that where the stresses are likely to be required, at the surface of the structure, is just the place where least information is available for averaging.

9.6 Methods of Stress Averaging

The simplest form of stress averaging consists of simply adding together corresponding stress components at a node and dividing by the number of elements that meet at the node and which are being included in the averaging process. In some cases this can lead to less accurate nodal stress estimates than other forms of averaging. Instead of taking the actual stress values at the node they can be weighted in some way so that the more important components have a greater significance in the averaging. The weighting used is usually based upon the element geometry and can be derived from the element Jacobian at the node. Typically the determinant of the Jacobian divided by the total element volume is used as the basis of the weighting function since this will emphasize those elements that have a large amount of material related to the node at the expense of those elements that are rather elongated. A variation on this is to use the included angle at the element node to provide the weighting function for the averaging. Again this can be found from the element Jacobian and serves to emphasize the effect of the elements with more material at the node.

Higher order elements often also require some form of averaging over the element. Most displacement elements use incomplete polynomials in their shape functions and the high order incomplete terms can 'contaminate' the stresses. For example, an 8-noded quadrilateral has a parabolic interpolation function for displacements, and when this is differentiated the strains are going to be essentially linear. However, the strains will still vary parabolically in some directions and linearly in others, but since the parabolic (and higher order) components are incomplete they constitute an error and show up as parabolic 'ripples' on the computed stresses. These ripples are not smoothed by nodal averaging since all elements meeting at a node can be simultaneously over (under) estimating the stress values at the node and under (over) estimating the values at other points on the element. In such cases the stresses will oscillate within the element about the correct value (assuming that the mesh is fine enough). This behaviour is a consequence of the fact that the finite element method only guarantees a mean square convergence over the element and not a point wise convergence which is needed for correct nodal stress estimates. It is quite possible to have a very refined mesh which has ensured convergence in the mean but which still has considerable errors in the nodal stress estimates at a point. In passing, it is worth noting that as the mesh is refined the stresses at a point need not converge monotonically as the displacement will. In such cases it is advantageous to be able to carry out some form of spatial averaging over the element. Typically for the 8-noded quadrilateral the actual stresses which are computed are replaced by an equivalent bi-linear interpolation to give the spatial average, where the stress interpolation is chosen to have the same average values as the nodal values which are directly computed. This then serves to smooth the stresses and removes the 'ripples'. A more common method of doing this in practice is to compute the stresses at a number of internal points within the element and to extrapolate from these points to the nodes. Typically for the 8-noded quadrilateral the stresses are extrapolated from the set of (2×2) 'reduced' Gauss integration points. Since the extrapolation is then only from four points it will be bi-linear and again will serve to smooth the stresses. This form of stress smoothing only works for a reduced set of integration points. A full integration of the 8-node quadrilateral demands that a (3×3) set of Gauss points are used. If the stresses are extrapolated from this full set then the extrapolation will be bi-quadratic and spurious parabolic 'ripples' will occur in the computed stresses. Similar arguments can be applied to other high order elements but it should be pointed out that few other elements can use reduced integration as advantageously as the 8-node quad or 20-node brick and it is sometimes necessary to use a full set of Gauss points for the element integrations, and a reduced set with extrapolation for the stress recovery in order to achieve the spatial stress smoothing. The Gauss points are not unique and the stresses at these points need not be more accurate than at any other point. The use of a reduced set of integration points tends to work both because of the position of the Gauss points and because it gives a bi-linear fit which smoothes out the second order error ripple. A more precise smoothing process involves a least square surface fit to the nodal stress values since this is in line with the least square convergence behaviour of the finite element method.

In the above discussion attention has been concentrated upon averaging the stresses within elements or across nodes. There are other forms of averaging that are possible which takes place over a group of elements (or possibly the complete structure) and involves some form of curve or surface fitting. Although this process can lead to very good stress estimates it is difficult to program for general shapes and is not usually found in programs.

9.7 Stress Presentation

After the stress recovery points have been chosen, there is then the problem of the form in which the stresses should be presented. Most FE systems give components of nodal or element stress in either global or local directions, depending upon the element type and the system being used. This presentation is not always the most convenient and sometimes, especially for elements with stresses in local coordinates, it can be rather confusing. It is useful to be able to obtain surface stresses, especially in the local surface directions, since this allows the satisfaction of stress-free boundaries to be assessed. When stresses are given in global coordinates these conditions can only be investigated where the free surface is parallel to a global axis. It is necessary to be able to calculate the principal stresses at a point since these give the absolute maximum stresses. The orientation of the principal stresses relative to the global axes should also be available. For a two dimensional stress field this is reducible to a single angle, usually relative to the global x-axis. For three dimensional stress systems the picture is very much more complicated since the principal stress directions will now be orientated along three local vectors making their presentation and understanding very much more difficult. Care must be taken when drawing stress contours using principal stresses since it is quite easy to make mistakes when two principal stresses become equal and follow the wrong one and so produce a discontinuity in the resulting contour. It is also easy to contour the wrong components across elements and obtain contours with a 'herringbone' look to them as the contours change direction discontinuously across elements.

In assessing stresses it is common to use a single equivalent stress for the purposes of assessment. This is necessary because it is very difficult to establish reliable stress failure criterion for multi-axial stress systems, partly because of the complexity of the resulting experiments and partly because of the difficulty of finding meaningful relationships between all of the parameters. The experimental problems are formidable for even

uniaxial experiments and a wide range of scatter is obtained on failure results. However, if the failure criteria is based upon uniaxial experiments it is then necessary to convert the multi-axial FE results into a single equivalent stress. There are many established ways of doing this and their definition will depend upon the mode of failure that the material undergoes. Most metals tend to fail by flowing in some fashion, either through plastic flow or through creep. This observation leads to the Tresca equivalent stress which is essentially saying that if any component of the stress exceeds a set value then the material will flow. A mathematically more refined version of this is the Von Mises equivalent stress. This uses two premises:
1. The material will not flow (fail) under conditions of hydrostatic stress.
2. The failure criteria must be independent of the axes that are used to define the stresses.

To satisfy the first condition, the hydrostatic (pressure) stress component is subtracted from the stresses before they are assessed. This gives the deviatoric components of stress. There are various ways of incorporating the second condition since three invariant stress functions can be defined. The first of these is the hydrostatic stress. The second stress invariant is used to give the Von Mises equivalent stress. There is a close relationship between the Tresca and the Von Mises equivalent stresses and there is no firm experimental evidence to support either as the definitive equivalent stress. The Tresca criteria is more conservative than Von Mises but the latter, being a continuous function is easier to handle within a computer program.

Some metals and alloys are more brittle than ductile and another criteria such as the maximum principle tensile stress is more relevant. The prediction of crack propagation, residual strength and fatigue life is usually based upon a stress intensity factor but many systems now have special facilities for calculating this. In the analysis of crack propagation there are various J-integral methods where the stresses are integrated around a closed contour around the crack tip to find the stress intensity factor. There are also other techniques such as the virtual crack extension methods to calculate the same quantity.

Materials other than metals can fail in modes that do not involve flow. Concrete, soils and other granular material are held together by frictional forces and they fail by sliding. In such cases it is important to include the effect of the stress normal to the failure plane in the failure criteria since the magnitude of the force that causes a frictional failure will vary with this normal stress. Two main criteria have been developed for frictional failures, the Mohr–Coulomb and the Drucker–Prager equivalent stresses. These are essentially developments of the Tresca and the Von Mises equivalent stresses to incorporate the effect of the normal stress. The failure criteria associated with friction failures are very much less well defined than the flow failures in metals.

There are also a large range of other failure modes associated with structures. The modern development of composite material has led to the investigation of a range of possible failure modes. These materials are layered which means that the invariant property of the Von Mises equivalent stress is no longer relevant since the failure will be directionally dependent. It also tends to mean that the individual stress components are much more likely to cause failure and single equivalent stress is not likely to be useful. There can be difficulties associated with using the FE method to predict such failures. Very often the actual stress that is causing the failure is not calculated, instead some smeared, average value is computed and there are problems in recovering the required component. Even worse from the finite element point of view is that it is not necessarily the largest stress that causes the failure. It was stated previously in this chapter that the error in any stress is the same for all stresses at about, typically, 5% of the maximum stress. This means that if a stress component with a magnitude an order less than the largest stress is causing the failure then the error in this stress can be 50% of its actual value. Obviously this is of little use for predicting failure.

9.8 Element Strains

The strain distribution throughout the structure is probably not required so often as the stresses but there are occasions, associated with the assessment of rupture and other forms of failure where a knowledge of the strains is important. When the displacement form of the finite element method is used the strains are always calculated as a step in the stress calculation and are, therefore, in theory available to the user. Whether they can be printed out then depends upon the FE system that is being used and the facilities that are available within it. All of the comments relating to the calculation of stresses that were made in the previous section are equally relevant to the strains. In addition it must be recognized that there can be various components to the total strains that are not present for the stresses. The total strain is a geometrical relationship obtained by differentiating the displacements. This will be the sum of the elastic strain, the thermal strain and any other initial strain.

The user should check any failure criteria that is being used that involves strains to ascertain whether they are total or just elastic strains. It is very convenient if FE systems allow the different components to be printed as this will give complete flexibility to the user as to how the strain information is subsequently used. In passing it is worth noting that ill-conditioning errors can arise in the stress recovery when the thermal or other initial strains are subtracted from the total strains to obtain the elastic strains that are used to compute the stresses. It is often the case that the thermal strains and the total strains are very nearly equal and if the elastic stress recovery is not carried out in double precision (that is at least 60 bit arithmetic) then significant rounding errors can occur giving the wrong stress recovery.

9.9 Calculation of Reaction Forces

Another piece of information that is used for assessing the behaviour of the structure is the support reaction forces. These are used to design attachments that hold the structure. The idea of reaction forces can be generalised to include inter-element forces, especially where the strength of internal structural joints have to be considered. FE programs should have facilities, not only for calculating reaction forces, but also to be able to print out individual element forces (stress resultants). These can then be used for local joint detailing, provided that they are used with care. If a series of nodes on one element are supported then the resulting reaction forces will in fact be a kinematically equivalent nodal force representation of a boundary stress field. If higher order elements are being used then these equivalent forces can vary wildly even for a constant stress field. Typically, a 20-node brick element under a uniform face pressure will have nodal forces that not only vary in magnitude but they also vary in sign. In such cases individual reaction forces can be very misleading and it is usually more meaningful to average the reaction forces over an edge or face of an element, typically by computing the end load and the bending moment on the element face or edge. The variation of the forces leads to many problems where users are trying to model contact problems by looking at the magnitude of reaction forces across the contact. For example if the force is in tension then it is assumed that the contact has broken and the freedom is released. Unfortunately the reaction force can be in tension but the contact pressure under the node can still be in compression. This leads to a very unreliable calculation procedure and such uncoupling methods should be used with care. In this situation low order elements tend to be better since the equivalent loads are more

A FINITE ELEMENT PRIMER

representative of the contact stresses. It might be felt that a more realistic model is obtained if a bar element is included to represent a bolt and the force in this element used to determine opening and closing of a flange rather than reaction forces. Unfortunately, the forces in these bolting elements are very likely to exhibit exactly the same behaviour as the reaction forces. As the stiffness of the bolt is increased then the closer the bolt force comes to the kinematically equivalent reaction force.

The interface pressures between two surfaces can be found from a subsidiary finite element calculation using the calculated reaction forces $\mathbf{R}$. For the edge of an element the interface pressure P_g can be interpolated using the element shape function, $\mathbf{N}$, along the edge and the element nodal interface pressure $\mathbf{P}_{Ng}$ as

$$P_g = \mathbf{N}\mathbf{P}_{Ng}.$$

Using the principle of virtual displacements the equivalent element nodal forces are

$$\mathbf{R}_g = \int_s \mathbf{N}^t \mathbf{N} \mathbf{P}_{Ng}\, ds = \mathbf{A}_g \mathbf{P}_{Ng}.$$

The nodal interface pressures can be assembled exactly as for the displacements so that the reaction forces are

$$\mathbf{R} = \sum_g \mathbf{a}_g^t \mathbf{A}_g \mathbf{a}_g \mathbf{P}_N = \mathbf{B}\mathbf{P}_N$$

where $\mathbf{P}_N$ are the assembled nodal interface pressures. These can then be found by solving

$$\mathbf{P}_N = \mathbf{B}^{-1}\mathbf{R}.$$

The interface pressures are physically meaningful whereas the element reaction forces are not.

9.10 Graphical Presentation of Results

It is very important that a user has a range of graphical facilities available for presenting the results of an analysis. For any real structure there is an enormous volume of data available in the results and a good deal of this has to be assimilated. There are two basic functions that the graphical output can satisfy. The obvious one is for the user to get a pictorial presentation of how the structure is acting under a given loading in order to ensure that important components of the response are correctly assessed. The second function of the graphical output is to allow the user to add a measure of quality assurance to the results by investigating the effects of the finite element modelling and to satisfy himself that the results are in fact acceptable. The features that are required of an output graphical processor are:

a. Deformed shape plotting, either for the complete structure, for the structural outline or with hidden line removal.
b. Contour plotting of any stress component on the surface of the structure and, for solid models, stress contours on internal sections.
c. Plots of graphs of stresses along lines on or within the structure. These stresses should be in either global coordinates or in the local direction along the line.
d. The facility to window in on sections of the mesh or on groups of elements. This should be independent of how data was generated since it is usually only when the results are being plotted that the user realizes what he wants to see. It is unfair to expect him to predict what combinations of views he is likely to want when the mesh is being generated. Contours should be scaled to values of the elements in the window rather than for global values. If this is not done then it can be difficult to obtain contours in the window if there is a stress concentration at some other point that is not displayed.
e. The ability to delete extraneous data from the plot so that the effects of interest can be viewed without them being confused by irrelevant data. This often arises by a window command plotting everything in the window without the user being able to plot only a slice within it.
f. The user should be able to plot the raw data from the finite element program, typically element stresses before averaging. This is especially important for the user to be able to assess the 'goodness' of the results and to confirm that the mesh is sufficiently fine. If only average values are available for plotting then a great deal of useful information is not made available to the user.

This last point illustrates an important point regarding graphical output. The emphasis tends to be on presenting 'pretty' pictures that are suitable for reports. While this is an obviously important requirement it should not be an overriding objective. The user should be able to produce a graphical representation of any data, including that which is not relevant to or suitable for any report. This form of plot is usually more informative for error checking purposes.

9.11 Using the Results to Refine the Mesh

Once the structure has been analysed and the results assessed then the analyst has a great deal of information that he can use to build up his knowledge base as to what constitutes a good mesh for that type of problem. Attempts have been made to formalise this information into post-processing programs that automatically generate better meshes for the same number of elements. The basic algorithm in most of these programs use the strain energy density as the function used for the optimization. This is easily calculated at any point as

$$SED = \tfrac{1}{2}\varepsilon^t \sigma.$$

In one class of methods the user then specifies lines of nodes, where the lines must include at least the boundaries of the mesh but it can also include other internal lines. The mesh refinement algorithm then redefines the position of the nodes along these lines such that change in strain energy density is constant from node to node. Having obtained the new node positions along the specified lines the algorithm then fills in the enclosed areas and volumes with a new mesh. It is this stage that provides the most difficulties. The process is easiest if triangular and tetrahedron elements are used but these are generally the least accurate elements. The automatic refinement is more difficult for rectangular and higher order elements and the current state of the art usually requires manual editing of the generated mesh before it can be used. There are also complications where the structure is subjected to multiple load cases since each of these would require a different optimum mesh. The load cases can be weighted in some manner so that an optimum mesh for the combined set can be found. Another suggested method of mesh improvement does not alter the element density. Instead elements that are in regions of high rate of change of strain energy density are replaced by higher order elements where the increase in the order is obtained by means of 'bubble functions' or by adding nodes to edges that do not connect to other elements. This approach has the distinct advantage of not requiring a change in the mesh but it does add extra equations to the stiffness matrix and these destroy the banding of the matrix. Provided that the refinement requires only a few new equations this is not a significant problem. The process can also be organized to use the majority of the existing factorised stiffness matrix so that a complete re-analysis is not required. Other variations on this theme are also being developed, all with the aim of improving the solution for the minimum of recalculation. If these techniques can be developed they are likely to prove a powerful advance in the use of the finite element method.

The current state of the art is such that automatic mesh refinement is not a practical tool for the analyst. This is because of the need for the user to edit the generated mesh, the fact that the modified mesh is load case dependent and because of the cost of repeating the analysis. However, it is often useful for the analyst to at least obtain the new mesh density along the edges and on defined lines within the mesh since this provides a visual feedback as to what constitutes a better idealisation. This can than be used as a guide for mesh generation of similar geometries and load cases. There are also possibilities of using this approach to devise an algorithm that can assess the accuracy of the final solution and provide the user with a confidence factor for the results.

10. Dynamic Analysis

10.1 Introduction

In a dynamic analysis the effects of inertia forces are considered in the calculation. These are proportional to acceleration and their inclusion leads to an equation that has time-varying terms, giving a dynamic response. In order to define such a problem the minimum information that the user has to specify is the stiffness of the system and the inertia of the system. For a response calculation some initial conditions for displacement and velocity are also required. In addition any real system will contain damping that dissipates the vibrational energy and, probably, a time-varying set of loads known as the forcing function.

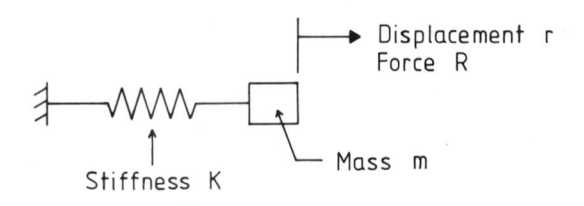

Figure 10.1

The simplest dynamic problem is that of the one degree of freedom system, typically the spring/mass structure shown in Figure 10.1 which has an equation of motion

$$m\ddot{r} + kr = R. \qquad (10.1)$$

This has a resonant, or natural frequency

$$w = \sqrt{k/m} \qquad (10.2)$$

DYNAMIC ANALYSIS

and if the structure is excited at this frequency then a very high amplitude response occurs. For this reason it is necessary to design structures such that the resonant and excitation frequencies are not close to each other, implying that the engineer must have some knowledge of the structure's resonant frequencies. If the system of Figure 10.1 is given some initial disturbance then it will vibrate at this natural frequency. Any real structure will contain damping, which is usually assumed to be viscous (that is the damping force is proportional to velocity), so that the equation of motion becomes

$$m\ddot{r} + c\dot{r} + kr = R. \qquad (10.3)$$

This is often written in the non-dimensional form

$$\ddot{r} + 2w\xi\dot{r} + w^2 r = \frac{1}{m} R \qquad (10.4)$$

where the various terms in the equation are called

undamped natural frequency $\quad w = \sqrt{k/m} \qquad$ (10.5a)

damping factor $\quad \xi = c/c_0 \qquad$ (10.5b)

critical damping $\quad c_0 = 2\sqrt{km} \qquad$ (10.5c)

damped natural frequency $\quad \beta = w(1-\xi^2)^{1/2} \qquad$ (10.5d)

decay factor $\quad \alpha = w\xi. \qquad$ (10.5e)

With viscous damping the structure resonates at a slightly different frequency, called the damped natural frequency, β. This is also termed the resonant frequency for a single degree of freedom system and is the frequency at which the maximum response occurs when it is excited by an harmonic force. The majority of structures are lightly damped with the damping factor $\xi < 0.1$. In this case it will be seen that the damped and the undamped natural frequencies are very nearly equal. If the system represented by equation (10.3) is given an initial starting displacement r_0 and has no forcing function then the resulting equation of free vibration is

$$r(t) = r_0 e^{-\alpha t}\left(\frac{\alpha}{\beta}\sin \beta t + \cos \beta t\right). \qquad (10.6)$$

It will be seen that this response dies away with time at a rate defined by the decay factor, α.

Real structures are rarely simple single degree of freedom systems and the multi-degree of freedom equations must be formed. This is most conveniently done in matrix form and the equation of motion that is usually solved is

$$\mathbf{M\ddot{r}} + \mathbf{C\dot{r}} + \mathbf{Kr} = \mathbf{R}(t) \qquad (10.7)$$

where $\mathbf{M}$ is the mass matrix, $\mathbf{C}$ the damping matrix, $\mathbf{K}$ the stiffness matrix and all of these are symmetric. $\mathbf{R}(t)$ is the vector of forcing functions and $\mathbf{r}$ is the resulting displacement response. The mass and stiffness matrices are considered in more detail in section 10.3 but they can always be formed in finite element terms once the element shape function has been assumed. The damping matrix is discussed in section 10.6. For the various reasons discussed in that section this matrix is rarely formed from first principles and, instead, some approximate distribution of the damping is usually assumed. In forming this equation it has been assumed that the kinetic energy is only a function of velocity and not of displacement. If this is not the case then equation (10.7) is probably not valid. In practice this usually arises where the structure is rotating about some axis, which gives rise to centrifugal and gyroscopic forces. The centrifugal forces can be included in the stiffness but any gyroscopic effects require a skew-symmetric 'damping' matrix and most general purpose packages cannot handle such matrices.

As a starting point consider the dynamic analysis of the simpler undamped free vibration equation

$$\mathbf{M\ddot{r}} + \mathbf{Kr} = \mathbf{0}. \qquad (10.8)$$

This defines the essential dynamic characteristics of the system and serves to give the user an insight into its response. The inclusion of the inertia terms give rise to considerable complications so far as the form of solution and the behaviour of the system is concerned. With a static analysis it is usually possible to visualise how the structure will behave under a given load. This is no longer true with a dynamic analysis and the user requires extra information in order to be able to interpret and assess the results. The undamped free vibration equation can be used to find the resonant frequencies and mode shapes of the system and a knowledge of these is invaluable for understanding the dynamic structural response. It is also the basic linking point between theory and experiment for dynamics since, for linear systems, the dynamics of the structure are fully defined by the resonant frequencies and the mode shapes and these can be both calculated and measured.

10.2 Undamped Free Vibrations – The Eigenvalue Problem

Equation (10.8) has a solution which can be written in the form

$$\mathbf{r} = \boldsymbol{\phi}_j e^{iw_j t}. \tag{10.9}$$

Substituting this into the equation of motion leads to the eigenvalue problem

$$\mathbf{K}\boldsymbol{\phi}_j = \lambda_j \mathbf{M}\boldsymbol{\phi}_j; \quad \lambda_j = w_j^2 \tag{10.10}$$

where λ_j is the jth eigenvalue and the column matrix $\boldsymbol{\phi}_j$ is the corresponding jth eigenvector. w_j is called the jth resonant frequency of the system and these are the frequencies that the structure responds most strongly to. If the structure has n degrees of freedom (there are n equations of motion) then there will be n eigenvalues and vectors. So far as the response calculation is concerned not all of these contribute and there is no need to find all n eigenvectors. Usually it is only the vibration modes that have the lowest eigenvalues that are important in the response and hence only the low modes need be found. The number of modes that are significant in a response is discussed in section 10.8. The eigenvalues are important physically because they define those frequencies at which the structure wants to vibrate, that is those frequencies that should be avoided in the forcing function. The eigenvectors are important mathematically because they have the property of defining a coordinate transformation that simultaneously orthogonalises the stiffness and the mass matrix. This means that

$$\boldsymbol{\phi}_i^t \mathbf{K} \boldsymbol{\phi}_j = 0 \tag{10.11a}$$

$$\boldsymbol{\phi}_i^t \mathbf{M} \boldsymbol{\phi}_j = 0 \tag{10.11b}$$

and if the m eigenvectors that have been computed are combined into a single matrix of the form

$$\boldsymbol{\phi} = [\boldsymbol{\phi}_1, \boldsymbol{\phi}_2, \boldsymbol{\phi}_3, \ldots, \boldsymbol{\phi}_m] \tag{10.12}$$

then

$$\phi^t K \phi = k \qquad (10.13a)$$

$$\phi^t M \phi = m \qquad (10.13b)$$

where **k** is the diagonal generalised stiffness matrix and **m** is the diagonal generalised mass matrix.

The eigenvalues and vectors are so important both theoretically and experimentally that they have been given various names. The eigenvalues are known as characteristic values, latent roots and the square root of the eigenvalue is the structural resonant frequency or natural frequency. The eigenvectors are called variously the characteristic vectors, the latent vectors, normal modes and mode shapes. All of these terms tend to be used synonymously. The eigenvectors can be presented in various ways. It will be seen from equation (10.10) that, as the vector occurs on both sides of the equation, it can be multiplied by an arbitrary constant. This multiplication of the vector by a constant is called normalisation. There are two main methods for normalising the vector, the engineering form, where the largest term in the vector is scaled to be unity and, secondly, the mathematical form, where the vector is scaled to make the generalised mass unity. The engineering normalisation can be used for external presentation (printing) of the vector and the mathematical normalisation is used internally for the response calculation. However, there is no unique method of normalising the eigenvectors and when mode shapes are compared between different programs or with experiments they usually have to be re-normalised to allow a direct comparison. Even within a given program a slight change can cause the eigenvectors to change sign since difference in the rounding process can cause a different term to be chosen as the normalising factor. Such an effect should not be considered as an error or even an inconsistency since it is only a manifestation of the arbitrariness within equation (10.10).

The eigenvalues are usually distinct in value and the resulting eigenvectors are unique, apart from the arbitrary normalising constant. If the structure has any symmetries there is a possibility of two or more eigenvalues having the same value. In this case the eigenvectors are not unique since linear combinations of vectors that have that same eigenvalue also satisfy equation (10.10). The user should be aware of this when comparing vectors between programs. If the structure does have symmetries then the eigenvectors will be either symmetric or anti-symmetric. In such cases advantage should be taken of the structural symmetries to solve for

DYNAMIC ANALYSIS

symmetric and anti-symmetric vectors separately since, this not only avoids linear combinations of the vectors being found, but it also makes the solution process cheaper. For structures that include bending effects and thus have both rotational and translational degrees of freedom, then the normalisation of the vector becomes dependent upon the units of length that are used since the ratio of translation to rotation displacements have units of length.

There is a large variety of methods for finding eigenvalues and eigenvectors and all of these have been used in various programs. For a static analysis the development of the finite element method has led to a small number of solution techniques being generally agreed as being best suited to the form of equations that arise and most programs use either a frontal or a Cholesky solution. There is no consensus as to which are the best methods for finding eigenvalues and vectors and many programs themselves contain a variety of methods. The analyst then has the problem of choosing which one to use. The best choice depends upon the actual form of the equations being solved, their size, their sparsity and the number of eigenvalues and vectors that are required. As the user is often called upon to make a choice of which routine to use the salient features of the various classes of methods of eigenvalue extraction are now discussed. Only the general features are given, not the full mathematical details. The various methods for solving the eigenvalue problem can be classified into one of three types.

a. Power methods – These iterate on the equations using something of the form

$$\mathbf{r}'_{i+1} = (\mathbf{K} - g\mathbf{M})^{-1} \mathbf{M} \mathbf{r}_i; \quad \mathbf{r}_{i+1} = a\mathbf{r}'_{i+1} \qquad (10.14)$$

where g is some shift and a the normalising factor. It can be shown that this converges to the eigenvalue λ_i where $(\lambda_i - g)$ is closest to zero. By suitably choosing the shift, g, theoretically any eigenvalue can be found. The obvious problem here is that some knowledge of the eigenvalues of the system is required and, unless considerable care is taken, it is easy to miss values. The other problem with this approach is that the coefficient matrix $(\mathbf{K} - g\mathbf{M})$ that has to be solved changes as the value of g is altered, making the method inefficient for finding anything other than one or two eigenvectors. This form of the power method is known as inverse iteration and is often used in conjunction with other techniques that calculate eigenvalues accurately but which are not convenient for finding eigenvectors. There are other variations of the power method where the shift is taken as zero. The lowest eigenvalue and vector are then determined and the convergence is very rapid, usually within half a dozen or so iterations. The second eigenvalue and vector is then calculated by choosing a starting vector that is mass orthogonal to the first. The inevitable rounding errors in the calculations then loses this orthogonality and if it is not continually enforced it is found that the process again converges to the lowest eigenvalue and vector. A variation of this method is discussed later as the subspace or simultaneous vector iteration method. If a buckling problem is being solved, where only the lowest eigenvalue is of interest then the power method is probably the best method to use.

b. Transformation Methods – These utilise the fact that the vectors orthogonalise the coefficient matrices. The basic form of these transformation methods is to repeatedly carry out the product

$$\mathbf{A}_i = \mathbf{Y}_i^t \mathbf{A}_{i-1} \mathbf{Y}_i; \quad \mathbf{A}_0 = \mathbf{K}^{-1} \mathbf{M} \qquad (10.15)$$

where $\mathbf{Y}$ is a self-orthogonal matrix,

$$\mathbf{Y}^t = \mathbf{Y}^{-1}. \qquad (10.16)$$

This will eventually reduce $\mathbf{A}_i$ to a diagonal matrix, where the terms on the diagonal are the eigenvalues of the system. The eigenvectors can then be found from the products of the orthogonalisation transformations,

$$\boldsymbol{\phi} = \mathbf{Y}_i, \mathbf{Y}_{i-1}, \mathbf{Y}_{i-2}, \ldots, \mathbf{Y}_2, \mathbf{Y}_1. \qquad (10.17)$$

Variations arise within the transformation method from the choice of the orthogonal matrices $\mathbf{Y}$. The original technique in this class of methods was the method of Jacobi. In this case $\mathbf{Y}$ is composed of sine and cosine terms that are chosen to make successive off diagonal terms in $\mathbf{A}_i$ zero. When the next term is made zero the previous one becomes non zero so that a number of sweeps have to be made through the matrix before all of the off-diagonal terms are small enough to be considered zero. In practice five to ten sweeps are required for most matrices. Other transformation methods include the LR and the QR methods and the QR method is probably the most stable and reliable method available for finding the eigenvalues and vectors of an arbitrary matrix. For structural dynamics the fact that the stiffness and mass matrices are both symmetric and at least positive semi-definite means that any of the transformation methods are stable and reliable. The efficiency of both the LR and the QR methods are improved by first transforming the coefficient matrix $\mathbf{A}_0$ to a tri-diagonal form (that is a matrix that has a leading diagonal and one diagonal above and one below the leading diagonal that are non zero with all other terms in the matrix being zero). This tri-diagonal form can be obtained in a finite number of steps. The techniques for finding the tri-diagonal form includes methods due to Givens, Householder and Lanczos. Both the LR and the QR methods can be organised such that the tri-diagonal form is retained. When this is used in conjunction with a shifting strategy the convergence of the QR method is very rapid. Transformation methods are undoubtedly the most stable techniques for finding eigenvalues and eigenvectors but they all suffer the major drawback from the finite element point of view of destroying the banding of the equations. The finite element method is only viable as a solution procedure because of this banding which means that transformation methods are not generally suitable within finite element programs. However, they are extensively used as a part of most methods that do take advantage of the banding of the equations, as discussed in the next sub-section. They are also used directly where some form of the condensation process described in section 10.7 has been used to reduce the size of the dynamic problem. Any form of condensation will give rise to fully populated matrices. Transformations are especially efficient on those problems where the set of equations being transformed are small enough to be held in the high speed memory of the computer. They become inefficient when blocks of the equations have to be swapped continually from backing store, either directly or when using virtual memory. The maximum size of the equation set that should be used then depends upon the amount of high speed core available to the user, but this is likely to be somewhere between 100 and 300 equations.

c. Sparse Matrix Methods – Typical finite element models might have many thousands of equations and nowhere near this number of eigenvalues and vectors are ever required for a structural response calculation. In addition the stiffness and mass matrices that arise are symmetric and sparsely populated. Special eigenvalue extraction techniques have been developed to take advantage of these properties and are such that only a small number of the lowest eigenvalues and the corresponding eigenvectors are calculated. The first method of this type that was developed is the subspace (or simultaneous) vector iteration method. This is essentially the power method where a set of m vectors are iterated simultaneously. Care is taken to ensure that the

A FINITE ELEMENT PRIMER

relevant orthogonality properties are maintained at all times. The method converges to the lowest m eigenvalues. In practice $m+b$ vectors are used for the iteration vectors where there are an extra b 'guard' vectors that are discarded at the end of the iterations. The use of these guard vectors serves to speed up the rate of convergence of the initial m vectors. Some programs make an automatic choice for the value of b. If the user has to specify the number of guard vectors then a value of three to six is normally sufficient. There are two main problems with the practical application of subspace iteration. Firstly the number of vectors, m, must be chosen before the process is started and often the user does not have sufficient information at this stage to decide upon the number that will be required. The cost of the method increases rapidly with the number of vectors for two reasons. Having to continually enforce orthogonality causes the computing time to rise rapidly with the number of vectors and if the size of the array used to hold the set of vectors exceeds the size of the high speed core then the resulting disc transfers slows the method considerably. The second difficulty with the method is that there is no inherent guarantee that it will converge to the lowest m vectors and, at times, it will miss some vectors. This depends upon the choice of the starting vectors and, apart from specifying the number of these, the user usually has little control over the choice of the starting vectors. It is always advisable to use a Sturm sequence check to confirm that the full set of lowest vectors have been found.

A more recent method that has appeared is the development of the Lanczos method for large sets of equations. In this a transformation matrix is constructed to transform the mass matrix to a triple diagonal form and the stiffness to a unit matrix. A fundamental feature of the method is that these transformed matrices can be formed directly, without the need to actually carry out the intermediate transformation products. Only the array space for a fixed number of vectors is required (typically five but the actual number depends upon the implementation of the method) irrespective of the number of eigenvalues which are to be found. As the terms in the tri-diagonal matrix are formed the QR method is used to calculate the eigenvalues of the resulting tri-diagonal set. It is found that these converge quite rapidly to the lowest eigenvalues. One problem with the method is that, as convergence is achieved, then it is necessary to enforce the orthogonality of the new columns of the transformation matrix against the converged eigenvectors. Often convergence is enforced against all vectors whether converged or not. The Lanczos method allows any number of vectors to be found and the process can be stopped either when a specific number of vectors have been found or when all of the eigenvalues up to a given cut off value are found. It is also easy to arrange the process so that it can be restarted to allow more vectors to be calculated. As with the subspace iteration method there is a possibility of vectors being missed. Tests tend to indicate that the Lanczos method is rather more efficient than subspace iteration unless only the first one or two eigenvalues and vectors are required.

Associated with these eigenvalue extraction methods is the Sturm sequence check. This is a process whereby the number of eigenvalues having values below a given estimate can be found, without having to determine the intermediate values. Each estimate requires a matrix factorisation which makes it expensive for anything other than one or two estimates. It can be used before the eigenvalue extraction to determine the number of modes below a given value, allowing the user to specify this as the terminating value for either subspace iteration or the Lanczos method. It is also useful as a pre-extraction check, especially on a highly detailed model, to find if there are a large number of low frequency secondary component modes (typically say local floor modes of a building where the overall response is required) which would make the extraction process very expensive to carry out. If this is the case then the structure has to be re-modelled. It is also useful for detecting silly mistakes with units, especially the mass units, since this will show up either as a very large number or no low frequency modes. The Sturm sequence check can also be used after the eigenvalue extraction to confirm that all of the modes within a given subspace have in fact been found. A Sturm sequence can be used as a check because it is very stable numerically. The method consists of counting the number of sign changes of the determinants of the principal minors of the matrix $(\mathbf{K} - g\mathbf{M})$. Its inherent stability arises because only signs of numbers are involved, not their values. The values of the intermediate determinants can be grossly in error but the Sturm sequence check is self correcting.

Some care has to be exercised by the analyst if the structure is not supported in some directions (it is free-free) or if it contains any mechanisms. In such cases the stiffness matrix will be singular and any solution method that involves solving a set of equations defined by the stiffness matrix will be very prone to ill-conditioning. The user has little control over this situation and if an unsupported structure is to be analysed it is suggested that the eigenvalue extraction process that is to be used is tested first on a small problem where the stiffness matrix is not supported. Within programs there are two main strategies that are employed to overcome this problem. The simplest is the shift technique used in the power method, except that the shift is taken in the form $(\mathbf{K}+g\mathbf{M})$. This matrix will be non-singular and can be used in place of the stiffness matrix in the solution process. The main disadvantage with this method is the choice of the shift value. If it is too small then the equations will still be very ill-conditioned but if it is too large then the convergence to the eigenvalues and vectors will be very slow. If the user has a choice for the shift and if he has some idea of the value of the first natural frequency then the shift should be of the order one tenth to one half this. The second technique is to repeatedly orthogonalise the estimates for the eigenvectors against the rigid body modes as the solution process progresses. The user generally has no control over this.

10.3 Modelling Considerations for a Dynamic Analysis

Obviously the requirements of modelling for a dynamic analysis revolve around obtaining the correct vibrational behaviour for the structure. A minimum requirement is that the low frequency behaviour of the structure is correctly represented such that the lower modes have the correct resonant frequencies and mode shapes. This will be a necessary requirement whether the response is being found via a modal analysis or by some other form of calculation. It can be shown that the eigenvalues are minimum values of the Rayleigh quotient R,

$$R = \frac{\mathbf{r}^t \mathbf{K} \mathbf{r}}{\mathbf{r}^t \mathbf{M} \mathbf{r}} \qquad (10.18)$$

where $\mathbf{r}$ are the displacements of the structure. This is minimised when R is an eigenvalue and $\mathbf{r}$ the corresponding vector. The top line of the quotient is proportional to the strain energy of the structure and the bottom line can be interpreted as the kinetic energy. Note that there was an exactly similar minimisation that represented the solution of the static problem except that the bottom line of the quotient in that case was the external work done. As with the static case the ways in which the Rayleigh quotient can be minimised gives a great insight into the requirements for modelling the structure. The lowest mode is the absolute minimum of the quotient and this is usually obtained by a mode shape which minimises the strain energy and maximises the kinetic energy. Provided that there is not a significant number low frequency local modes within the structure (and if there is the model should be changed) then this statement will be generally true for all of the lower modes. The

DYNAMIC ANALYSIS

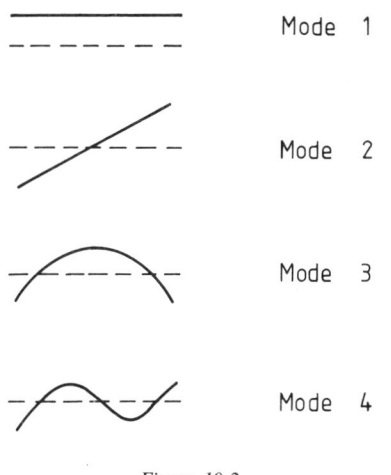

```
————————            Mode 1
— — — — — —

    ╱
———╱————            Mode 2
  ╱

   ╱‾‾‾╲
——╱————╲—           Mode 3
        ╲

—╱╲——╱╲—            Mode 4
    ╲╱
```

Figure 10.2

argument can be illustrated by considering the first few bending modes of an unsupported beam shown in Figure 10.2.

The first mode is a rigid body motion where the strain energy is zero and the kinetic energy is the total mass of the beam. The strain energy cannot be less than zero and the kinetic energy cannot be more than the total mass (assuming that the Engineers normalisation has been used for the vectors). The second mode is also a rigid body motion where the strain energy is still zero. In this case the kinetic energy is less than that for the rigid body translation since one point on the beam is stationary and does not contribute to the kinetic energy. The third mode is a half sine curve. This has strain energy which is proportional to the curvature of the beam. The kinetic energy must be less than that for the rigid body motions since there are now two stationary points on the beam. The fourth mode is a full sine curve where the curvature and hence the strain energy must be greater than that for the third mode. Similarly the kinetic energy is less since there are now three stationary points. This argument applies to each succeeding mode.

With these points in mind it will be seen that the structure should be modelled such that an accurate minimisation of the strain energy can be achieved and that the kinetic energy can be maximised. Maximising the kinetic energy essentially consists of summing together terms in the mass matrix in proportions defined by the values of the displacements. The process of summing the terms is not at all critical upon the distribution of terms within the mass matrix, the main requirement being that the correct proportions of the structural mass are associated with each node. This can be achieved by a diagonal mass matrix almost as well as by a more complicated banded (kinematically equivalent) mass matrix. A diagonal, or lumped, mass representation is often used, but such a representation is only valid for the low modes. If the user requires a better representation of the higher frequency response then the lumped mass form is progressively less accurate and the kinematically equivalent mass should be used.

The minimisation of the stiffness matrix is essentially a differencing process and, in this case, the distribution of the off-diagonal terms in the matrix is very important. The degree to which the stiffness has to be represented then depends upon what is required from the results. If only the resonant frequencies and mode shapes or the displacement response are required then a relatively coarse model will suffice. However, if the acceleration or stresses are required then a much finer mesh will have to be used. If stresses are being found then the same mesh that would be used for a static analysis is needed. In addition, in this case, care has to be taken as to how the stresses are recovered. This point is discussed in section 10.10.

All structures are designed for a purpose and rarely exist in isolation. Most carry non structural components. In some cases such non structural components are light compared to the structure whereas, in other cases, the non structural mass can be greater than that of the structure. Where the non structural mass is significant compared to the structure, typically more than 10%, then its distribution should be included in the mass idealisation. Very often the modelling of this non structural mass provides the analyst with considerably more problems than the actual structural idealisation. If the non structural mass is effectively rigid then it can be modelled as a point mass at its centroid, provided that the effects of both the translational and the rotational inertia are considered. In this case the analyst should use a lumped mass idealisation only if there is a structural node at the centre of gravity of the non structural mass. This sometimes requires the use of very stiff elements to position the centroidal nodes correctly. What must be avoided is the combination of modelling steps where rigid links are used to position a node at the centre of gravity of the mass and this is then followed by the user requesting a lumped diagonal mass matrix. This will cause the centre of gravity of the mass to be shifted to the point on the element that connects to the rigid link and can cause significant errors in the lowest frequencies. If rigid links are used

DYNAMIC ANALYSIS

for the purposes of mass idealisation then a kinematically equivalent mass matrix must be used. If the use of constraint equations (see chapter 12) is allowed in an eigenvalue extraction scheme (or the solution process that is being used) then they should be employed to model such idealisations in preference to other options.

When considering where nodes should be placed for the mass idealisation then the fact that, for the low modes, the kinetic energy has to be maximised means that the nodes should be positioned where the mass is large. For the stiffness idealisation the possible load paths that carry the inertia loads must be considered. If these load paths pass through regions of low stiffness then such regions should be modelled in detail since it is here that (relatively) large deflections can occur but, since the stiffness is low, the strain energy can be small. However, if the displacements are large the kinetic energy will be high thus giving the two conditions required to define a low frequency mode. In effect a dynamic analysis will always find any weakness in a structure and there is a good case to be made for always calculating the resonant frequencies of any structure even if only a static analysis is required since this will highlight any poor features in a structural design. Weak points in a structure can occur because part of the structure can bend or, more usually, because of structural joints. Unfortunately these are very difficult to model within a finite element analysis, usually requiring some form of estimated connection stiffness. This is illustrated by the simple structure shown in Figure 10.3, where the joint stiffnesses on a simple beam model were changed locally by an order of magnitude and it was found that the fundamental frequency changed by 30%.

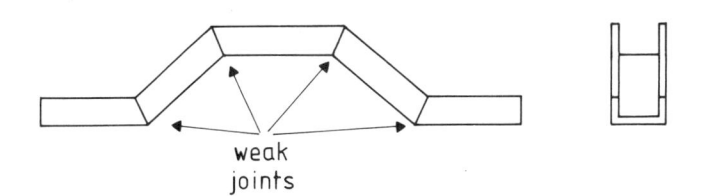

weak joints

Figure 10.3

Interestingly the mode shape did not change significantly and it was not possible to tell which was the weak model from the plots of the first two modes. Joint idealisation is less of a problem for structures that have been subject to a rigorous detailed design where the joints themselves have been considered, or where weak sections of the structure are short circuited by a stiff component. It is much more of a problem for those cases, typical of, say, the dynamic analysis of a domestic machine, where a series of components have been bolted together to form the machine, with no thought having been given to the structural design of the connections.

Nothing has yet been said regarding the modelling of damping. For most structural vibrations this is both small and ill-defined and it is not modelled in the same detail as the mass or the stiffness. It is discussed in section 10.6.

10.4 Forced Response

Generally it is necessary to find the dynamic response of the structure to some form of time-varying forcing function. The most efficient method of calculating this reponse then depends upon the actual form of the forcing function, and it is this which will dictate the subsequent solution strategy that should be adopted. The time variation of a force can be considered to lie in one of four categories, these being:

a. A periodic force input, typically the triangular wave shown in Figure 10.4. For such a forcing function the initial transients that occur when the force is first applied are not significant since the vast majority of the response will consist of a steady state vibration. The magnitude of this steady state behaviour is of interest for the purposes of fatigue and, possibly, acceleration level assessment.

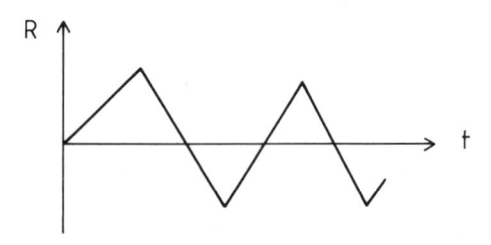

Figure 10.4

b. A transient force input, typically the pulse shown in Figure 10.5. In this case it is the transient response of the structure that is of interest, especially peak stresses, displacements and accelerations. The response tail that occurs after the peak will continue for a relatively long time if

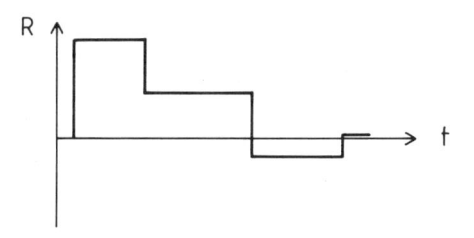

Figure 10.5

the structure is only lightly damped but the response in this tail will always be less than the peak and will not usually continue for a significant number of cycles so that it need not be considered in fatigue calculations.

c. A stationary random excitation, typically the random signal shown in Figure 10.6. This is the random equivalent to the periodic deterministic force input of case (a) and it is necessary to determine the stationary random response for fatigue studies.

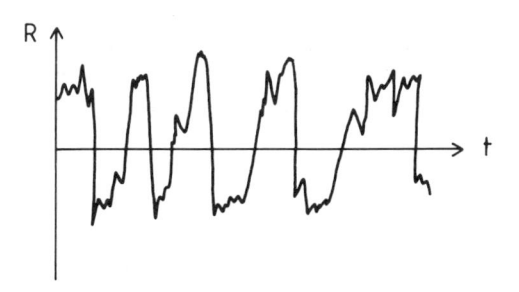

Figure 10.6

d. A non-stationary random excitation, typically the earthquake signal shown in Figure 10.7. This is the random equivalent of the deterministic transient input (b) and estimates of the probable peak stresses and displacements are required.

There are various different methods for calculating the structure's dynamic response and each of these tends to be efficient for finding the response to a given class of force input but is inefficient for other classes of forcing function. Typically, a method that calculates the response in terms of a

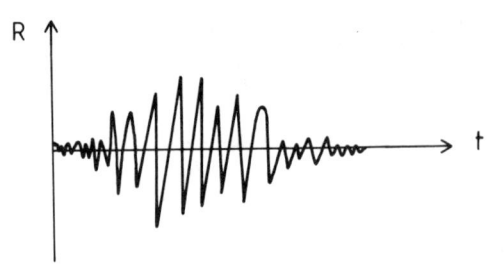

Figure 10.7

time history is well suited to finding the transient behaviour but is an expensive way of finding the steady state since it would require cycling through something of the order of 100 cycles before this is established.

10.5 Methods for Calculating the Forced Response

There are so many different methods available for a dynamic response calculation that the analyst must have a good knowledge of the advantages and disadvantages of every method that is available within the system he is using. The availability of various techniques can often dictate which program should be used for a given problem. It is possible to obtain the dynamic solution in either the time domain or the frequency domain.

For a solution in the time domain the input is specified as a time history and the response, be it displacements, velocities, accelerations or stresses, is also calculated as a time history. Generally these methods are started by defining some initial conditions for displacement and velocity and the time history of the input. The solution then takes small steps in time to calculate the time history of the response. A solution in the time domain is good for transient response calculations.

Alternatively the input can be defined in terms of its frequency content and the solution found in the frequency domain. In this case the steady state response is found directly and the initial conditions are irrelevant. The response is calculated at the same frequencies that are used to define the input. It is always possible to go from the time domain to the frequency domain via a Fourier transform and from the frequency domain to the time domain via an inverse Fourier transform, thus the two general solution methods are closely related.

DYNAMIC ANALYSIS

A solution in the time domain can be found either from a modal solution or from a direct step-by-step integration of the equations of motion. The modal solution takes the form

$$\mathbf{r}(t) = \int_0^t \mathbf{W}(t-\tau)\mathbf{R}(\tau)\,d\tau \qquad (10.19)$$

where $\mathbf{W}$ is called the impulse response matrix for the structure. This can be found in terms of the eigenvalues and eigenvectors of the structure. Equation (10.19) is variously called the convolution or the Duhamel integral. If the force input $\mathbf{R}(\tau)$ can be defined in terms of a continuous time function then the integration can be carried out explicitly to obtain the response at any time. A general force input can be idealised as a series of straight line segments and again this can be integrated explicitly to obtain the response. Since the integration is not carried out numerically there is no problem associated with the length of the time step and the same solution is obtained for any step length. The cost of a modal solution is all incurred in the calculation of the eigenvalues and eigenvectors. Once these have been found the calculation of the response history is then relatively cheap provided that the method has been programmed efficiently and the analyst uses the program intelligently.

Alternatively, the solution in the time domain can be found by some form of direct step-by-step integration of the equations of motion. These solutions generally make an assumption as to how the acceleration varies over a small time step Δ. Most of the current methods assume a linear variation of acceleration. The displacement, velocity and acceleration are known at the start of the step and the forcing function is known at the start and the end of the step. The acceleration at the end of the step is then found by solving an equation of the form

$$\ddot{\mathbf{r}}_{t+\Delta} = \mathbf{A}^{-1} f(\mathbf{R}_{t+\Delta}\ \mathbf{R}_t\ \mathbf{r}_t\ \dot{\mathbf{r}}_t\ \ddot{\mathbf{r}}_t\ \mathbf{r}_{t+\Delta}\ \dot{\mathbf{r}}_{t+\Delta}) \qquad (10.20)$$

and the velocities and displacements found by integrating the assumed acceleration variation in the form

$$\dot{\mathbf{r}}_{t+\Delta} = a_1 \ddot{\mathbf{r}}_t + a_2 \ddot{\mathbf{r}}_{t+\Delta} + \dot{\mathbf{r}}_t \qquad (10.21a)$$

$$\mathbf{r}_{t+\Delta} = b_1 \ddot{\mathbf{r}}_t + b_2 \ddot{\mathbf{r}}_{t+\Delta} + \Delta\dot{\mathbf{r}}_t + \mathbf{r}_t \qquad (10.21b)$$

where the a's and b's in these equations arise from the assumed time variation of the acceleration. Knowing these values at the end of this step means that they can be used as starting values for the next and the solution procedes over the required time interval. There are two variations to the solution of equations (10.20) and (10.21). In the first method the acceleration found by solving equation (10.20) is only approximate and the corresponding estimates for the velocity and displacement at the end of the step, found from equation (10.21) are fed back into equation (10.20) and an iterative loop is established until the accelerations are the same between two iterations. This gives the acceleration at the end of the step. For this approach the coefficient matrix **A** is simply the structure's mass matrix and this is very often constant for all time, even for most non-linear problems, since these usually involve small strains and, for most elements, the mass matrix is invariant with rigid body rotations. The disadvantage of the method is that if the time step is made too large then the whole process is numerically unstable and the calculated response amplitude grows rapidly with time. The problem here is that the time step is related to the period of the highest eigenvalue of the structure which is likely to be many orders of magnitude shorter than the time period of the fundamental mode, implying that many hundreds of steps are required for one cycle of the structure's fundamental frequency. Such methods of solution are said to be conditionally stable and will only work if the time step is short in relation to the period of the highest natural frequency of the system being analysed. Typically, the step length has to be smaller than one quarter of this period. They are well suited to the solution of shock and impact problems where the initial response is all important.

In the second form of the step-by-step solution equation (10.21) is used to eliminate the velocity and displacement at the end of the step from equation (10.20). At the same time approximations are introduced into the integration of the assumed accelerations for the velocities and displacements in such a way that the solution can be made numerically stable, but not necessarily correct, for any value of the time step. Eliminating the end-of-step velocities and displacements from equation (10.20) means that the accelerations at the end of the step are found directly, without the need to iterate within a step. The two most common forms of this method are the Newmark-β and the Wilson-θ methods, although there are very many other methods of step-by-step integration described in the literature. The constants β and θ in the Newmark and Wilson methods are used to control the stability of the process. If $\beta > 0.25$ then the Newmark method is stable for any time step and if $\theta > 1.36$ the Wilson method is stable. This statement is only correct unconditionally if the system is linear. If a non-linear dynamic problem is being solved it is not possible to make general claims for unconditional stability of any step-by-step method. In some implementations of the step-by-step method other constants are made available to control the stability of the solution but these usually have relatively little effect. The stability of the methods arise from the approximate integration of the assumed accelerations which gives rise to extensions to the time period of the oscillations and to the introduction of artificial damping. For the Newmark method it can be shown that these effects introduce an error of less than 1% if there are 20 integration steps over a given period. It follows that, for the unconditionally stable methods, the user can choose an upper frequency for which he wants accurate results and then choose a time step which gives at least 20 response times within the period of this upper frequency.

Within the literature step-by-step integration methods are classified as being either explicit, where the solution at $(t+\Delta)$ is obtained by considering the equation of motion at time t, or implicit, where the solution at time $(t+\Delta)$ is found by considering the equation of motion at time $(t+\Delta)$. In practice this distinction is artificial and it can be shown that the Newmark and most other methods can be derived in various ways, some being implicit and some being explicit. This can give rise to a certain degree of confusion. So far as the user is concerned the more important distinction is between conditionally and unconditionally stable algorithms. Unconditionally stable algorithms allow large time steps to be used but they are only accurate for linear problems with a low frequency response. Conditionally stable algorithms are used for non-linear or high frequency problems. The distinction between conditional and unconditional stability is itself indistinct since most programs allow the user to vary the parameters used within the method so that the same algorithm can fit both classifications.

There are also two different response methods available for steady state solutions in the frequency domain. The force input is Fourier transformed into its frequency components, w, and a series of equations of the form

$$\mathbf{M\ddot{r}} + \mathbf{C\dot{r}} + \mathbf{Kr} = \mathbf{G}\,e^{iwt} \qquad (10.22)$$

have to be solved. The amplitude of the force input, $\mathbf{G}$, is constant and can be a complex number. The real part of this is the in-phase component of the force and the imaginary part the out-of-phase component, relative to some arbitrary basis. The solution will be of the form

$$\mathbf{r} = \mathbf{g}\,e^{iwt} \qquad (10.23)$$

where **g** is the displacement amplitude and again it is complex with in and out of phase components. Substituting this back into the equation of motion gives the response amplitude as

$$\mathbf{g} = (\mathbf{K} + iw\mathbf{C} - w^2\mathbf{M})^{-1}\mathbf{G} = \mathbf{H}(w)\mathbf{G}. \quad (10.24)$$

This can be repeated for each frequency component in the force input. The term, **H**(w), is called the dynamic flexibility matrix of the structure and is the Fourier transform of the impulse response matrix. Such a solution can be applied directly to find the response at each frequency without the need to calculate the eigenvalues and vectors of the structure. However, this has two major drawbacks. The direct solution requires solving a different set of equations for each frequency which makes it expensive for an input containing more than about six frequency components and, secondly, the equations become ill-conditioned when the excitation frequency approaches any resonant frequency. This is unfortunate since this is precisely where the response is highest and of most importance.

Both of these difficulties can be removed if a modal solution is employed to obtain the inverse involved when forming the dynamic flexibility. This takes the form

$$\mathbf{H}(w) = \phi(\mathbf{k} + iw\mathbf{c} - w^2\mathbf{m})^{-1}\phi^t \quad (10.25)$$

where now only the trivial inverse of a diagonal matrix is required and this is accurate no matter how close the excitation is to any resonant frequency. As with the time domain solution the main cost of the modal solution is incurred in the calculation of the eigenvalues and vectors and once these are known the actual response can then be found relatively cheaply. The solution given by equation (10.25) assumes that proportional viscous damping has been used (see section 10.6). More precise damping models can be used but the solution corresponding to equation (10.25) is rather more complex.

10.6 Damping Idealisation

So far a viscous damping matrix, **C**, has been implied or included in some equations in this chapter but the actual form of the damping has not been discussed in detail. There are some structures where a well defined damping mechanism exists. This can arise either because specific dampers are included in the structure or there is a specific natural damping

DYNAMIC ANALYSIS

mechanism. These include fluid flow past the structure, giving rise to a drag force, and pressure wave radiation into the medium surrounding the structure. On occasions this radiation damping can be a very efficient form of energy dissipation. For a radiating plane wave the viscous damping per unit surface area of the structure is ρc where ρ is the density and c the speed of sound of the medium into which the structure propagates this energy. Radiation into a solid medium gives a high value for this damping and accounts for most of the foundation damping of buildings. It is also a relatively efficient energy dissipation mechanism for structures surrounded by a liquid and, in some cases, sound radiation into the gas surrounding the structure is the dominant form of damping. The treatment of specific damping mechanisms within the response solution is discussed later in this sub-section. More usually there is no specific damping mechanism involved in the structure being analysed and damping arises from a wide range of effects, all of which are small but not necessarily viscous. Under such circumstances no attempt is made to model the damping from detailed material or structural properties and instead any convenient simple assumption regarding its value is used. There are two common assumptions used in practice, modal damping and proportional (or Rayleigh) damping, both of which can be used for modal solutions but only Rayleigh damping is useful for non-modal solutions. The magnitude of the damping is usually expressed as a percentage of the critical damping value. For a one degree of freedom system the equation of motion is, as discussed previously:

$$m\ddot{r} + C\dot{r} + kr = R \qquad (10.26)$$

which can be written in non-dimensional form as

$$\ddot{r} + 2w\xi\dot{r} + w^2 r = \frac{1}{m} R \qquad (10.27)$$

where ξ is the damping factor

$$\xi = c/c_0 \qquad c_0 = 2\sqrt{km} \qquad (10.28)$$

and c_0 is the critical damping value for the system (that is the value at which it just oscillates). With damping the resonant frequency of the structure becomes

$$\beta = w(1 - \xi^2)^{1/2} \qquad (10.29)$$

which is called the damped resonant frequency. For most structures the damping is only small, being some fraction of the critical value. This means that the forces arising from the damping are small compared to the stiffness and the inertia forces. Damping is only of major significance where the stiffness and inertia effects cancel, that is at the structural resonant frequencies. For all structures with damping less than about 10% of the critical value it will be seen from equation (10.29) that the presence of damping will not significantly alter the resonant frequency from the undamped value. Also, if the forcing function is such that it does not have any frequency content at any of the structural resonant frequencies then the actual value of the damping used will not have a great effect upon the results of the response calculation. If the structure is excited at its resonance then the response is controlled entirely by the damping. In most cases such a correspondence between resonant and excitation frequencies is to be avoided since it is precisely this situation that causes a dynamic failure and the structure is modified to shift its resonances. There are occasions where this situation cannot be avoided, typically a rotating machine being brought up to speed will pass through various resonances. In this case a transient calculation should be conducted since the time duration over which the coincidence of the frequencies occurs will control the response. This can be conducted with zero damping. There still remains a few occasions where none of these remedies are relevant, especially for random vibrations. In such cases the user must choose some estimate for the damping within the system.

The basis of all of the modal solutions is to use the eigenvectors to transform the equations of motion to the individual modes of vibration. At this stage it is possible to add damping to each mode. The damping is usually expressed in terms of percentage critical damping for each mode. Accepted values of damping factors for typical forms of construction are:

a. Continuous metal structures — 2% to 4% critical
b. Jointed (bolted) structures — 4% to 7% critical
c. Prestressed concrete structures — 2% to 5% critical
d. Reinforced concrete structures — 4% to 7% critical
e. Small diameter piping systems — 1% to 2% critical
f. Equipment and large diameter piping — 2% to 3% critical.

For modal damping the damping matrix, C, is not formed explicitly, instead the damping is included after the eigenvectors have been used to transform the equations of motion to the uncoupled single degree of freedom form. The other common form of damping that can be used for all methods of analysis is to assume that the damping matrix is

DYNAMIC ANALYSIS

proportional to a linear combination of mass and stiffness in the form

$$\mathbf{C} = \alpha \mathbf{M} + \beta \mathbf{K}. \tag{10.30}$$

The factors α and β are chosen to give the correct damping at two frequencies. For a specified damping factor ξ at a frequency w then

$$\xi = \frac{1}{2}\left(\frac{\alpha}{w} + \beta w\right) \tag{10.31}$$

and using this at two frequencies w_a and w_b with required damping factors ξ_a and ξ_b respectively gives

$$\alpha = 2 w_a w_b (\xi_b w_a - \xi_a w_b)/(w_a^2 - w_b^2) \tag{10.32a}$$

$$\beta = 2(\xi_a w_a - \xi_b w_b)/(w_a^2 - w_b^2). \tag{10.32b}$$

It might be expected that if the values of ξ_a and ξ_b are chosen to give the same damping factor ξ at the two different frequencies then the damping will be the same at all frequencies. Unfortunately this is not so and it can be shown that the actual distribution of damping with frequency follows the curve of Figure 10.8. The damping is less than ξ for frequencies between w_a and w_b and greater than ξ above these. Modal damping allows a much more precise control over the actual damping used than proportional damping does.

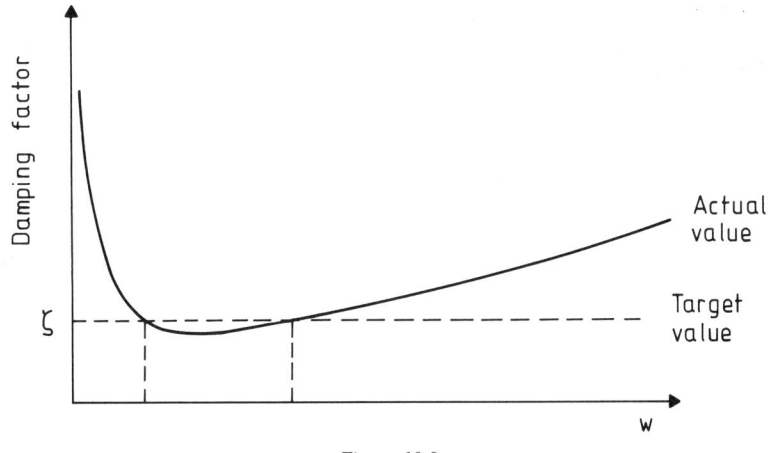

Figure 10.8

Where the specific damping mechanism is well defined and the resulting damping is greater than about 10% of critical then the actual damping matrix should be formed. This can be done by a standard application of the finite element method and can easily be incorporated into either the step-by-step numerical integration procedure or the direct calculation of the dynamic flexibility. It is not so easy to include it into a modal calculation since the eigenvectors that arise from the undamped eigenvalue problem will not orthogonalise the arbitrary damping matrix and they cannot be used to uncouple the equations into a set of single degree of freedom systems. Instead the equations of motion have to be re-written as

$$\begin{bmatrix} 0 & M \\ M & C \end{bmatrix} \begin{bmatrix} \ddot{r} \\ \dot{r} \end{bmatrix} + \begin{bmatrix} -M & 0 \\ 0 & K \end{bmatrix} \begin{bmatrix} \dot{r} \\ r \end{bmatrix} = \begin{bmatrix} 0 \\ R \end{bmatrix} \quad (10.33)$$

and the damped eigenvalue problem is

$$\begin{bmatrix} -M & 0 \\ 0 & K \end{bmatrix} \begin{bmatrix} \phi_{i1} \\ \phi_{i2} \end{bmatrix} = -\lambda_i \begin{bmatrix} 0 & M \\ M & C \end{bmatrix} \begin{bmatrix} \phi_{i1} \\ \phi_{i2} \end{bmatrix}. \quad (10.34)$$

This introduces considerable complications into the determination of the resulting eigenvalues and vectors, notably, the matrices are no longer positive definite and the eigenvalues and vectors can be (and will be for damping less than critical) complex conjugate pairs. Further the banding of the equations is destroyed and, although this can be reinstated it requires a non-standard assembly of the equations. Most systems do not provide facilities for solving such a damped eigenvalue problem but where they do a further complication arises. In expanding the equations to the $2n$ set of equation (10.33) it will be seen that both the displacements and velocities are solved for independently. However, these quantities are not independent in that the velocities are the time derivative of the displacements and hence they can be found by two entirely different calculation procedures. It can be shown that the only time that these two procedures agree is if the full set of $2n$ modes are used for the response calculation. With arbitrary damping in the equations then modal condensation (see section 10.7) cannot be used. Instead the equations must be condensed to a manageable size before the damped eigenvalue problem is solved. In passing it is worth noting here that if the structure is rotating about an axis and gyroscopic forces are present then their inclusion in the equations of motion leads to a skew symmetric coefficient matrix that multiplies the velocities in the same way that the damping matrix does. However, since the matrix is skew symmetric it never dissipates any energy but instead serves to transfer energy between modes. Such problems must be solved in the same way as the damped eigenvalue problem. This type of analysis is required for calculations of the whirling stability of shafts and rotors where a dynamic instability can be identified as where the real part of any eigenvalue goes to zero. The solution of dynamic problems of rotating machinery is very complicated, involving elastic stiffness, loading (centrifugal and other) stiffness and the choice of the analysis frame of reference (where it can be static or rotate with the machine). Such problems must be approached with care.

The most common assumption is that the damping is viscous and this is valid for most significant forms of damping. The other damping model that is used is structural or hysteretic damping. It has been found experimentally that internal energy dissipation within a material has the characteristic that, for a steady state cyclic response, the energy dissipated per cycle is independent of the frequency of excitation. For viscous damping the energy dissipation per cycle is proportional to frequency. Material damping is then idealised as viscous damping with the coefficient divided by the excitation frequency so that the equation of motion with material damping is

$$\mathbf{M}\ddot{\mathbf{r}} + \frac{1}{w}\mathbf{H}\dot{\mathbf{r}} + \mathbf{K}\mathbf{r} = \mathbf{G}\,e^{iwt}. \tag{10.35}$$

Since this requires a steady state input and a corresponding steady state response this can be written in the form

$$\mathbf{M}\ddot{\mathbf{r}} + (\mathbf{K} + i\mathbf{H})\mathbf{r} = \mathbf{G}\,e^{iwt} \tag{10.36}$$

where the stiffness matrix is now complex, the imaginary part being the material damping. In practice this is usually formed by replacing the static Young's modulus with a complex modulus of the form

$$E' = E(1 + i\eta) \tag{10.37}$$

where η is the material loss factor. This idealisation corresponds to replacing the non-linear hysteretic stress/strain curve of the material with an equivalent ellipse that encloses the same area. It must be emphasised that this idealisation is only valid for steady state response in the frequency domain. It must not be used in the time domain.

10.7 Condensation and Dynamic Substructuring

One problem with all forms of dynamic analysis is the cost in terms of computer resource. It is obviously convenient to use the same finite element model that was built for the static analysis but this often contains much more detail than the dynamic analysis requires, especially if the aim is only to find the lowest natural frequencies. There are various strategies that are available to the user which, if used intelligently, will reduce this cost but always at the expense of reducing the accuracy of the solution. In some cases this loss of accuracy is not significant but there are a few occasions where it can cause serious errors. It is unfortunate that such methods do not give the user any feedback as to the loss of accuracy that they engender. They must be used with care and only by analysts who fully understand the consequences of their application and who have the experience to assess the results realistically. There are two methods for reducing the size of the dynamic problem, condensation and dynamic substructuring. These two methods are closely related and some condensation techniques can be applied directly to substructuring. The n equations of motion to be solved are

$$\mathbf{M}\ddot{\mathbf{r}} + \mathbf{C}\dot{\mathbf{r}} + \mathbf{K}\mathbf{r} = \mathbf{R}(t). \tag{10.38}$$

These can be reduced to m equations, where $m \ll n$, by defining a coordinate transformation of the form

$$\mathbf{r} = \mathbf{T}\mathbf{s} \tag{10.39}$$

where the transformation matrix, $\mathbf{T}$, is of size $(m \times n)$. Applying this leads to a condensed set of equations of motion of the form

$$\mathbf{T}^t\mathbf{M}\mathbf{T}\ddot{\mathbf{s}} + \mathbf{T}^t\mathbf{C}\mathbf{T}\dot{\mathbf{s}} + \mathbf{T}^t\mathbf{K}\mathbf{T}\mathbf{s} = \mathbf{T}^t\mathbf{R} \tag{10.40}$$

or

$$\mathbf{M}_r\ddot{\mathbf{s}} + \mathbf{C}_r\dot{\mathbf{s}} + \mathbf{K}_r\mathbf{s} = \mathbf{S}'. \tag{10.41}$$

The problem is to then define the transformation matrix such that the accuracy required for the dynamic calculation is preserved, but with m as small as possible so that the cost is minimised. The most accurate transformation is to make $\mathbf{T}$ the first m eigenvectors of the system

$$\mathbf{T} = \boldsymbol{\phi}. \tag{10.42}$$

This gives a modal condensation and is the basis of the efficiency of all of the modal solution techniques. The orthogonality properties of the eigenvectors means that the condensed mass and stiffness matrices are diagonal.

Obviously modal condensation is of no use in reducing the size of the eigenvalue problem and some other method is required for forming the transformation matrix. Following the argument that the lowest eigenvalues are the minima of the Rayleigh quotient then these can be obtained by preserving the accuracy of the stiffness representation but approximating the mass distribution. This is done by defining the transformation matrix, **T**, in terms of the static behaviour of the structure alone. A set of displacements are chosen either by the user or automatically by the program and are designated the 'master' freedoms, r_m. The remaining freedoms, r_s, are termed the 'slave' freedoms. In the Guyan reduction method unit values are given to each master freedom in turn, with the other master freedoms held fixed and the resulting slave freedoms are found. This is done by solving the static equations

$$\begin{bmatrix} \mathbf{K}_{mm} & \mathbf{K}_{ms} \\ \mathbf{K}_{sm} & \mathbf{K}_{ss} \end{bmatrix} \begin{bmatrix} \mathbf{r}_m \\ \mathbf{r}_s \end{bmatrix} = \begin{bmatrix} \mathbf{R}_m \\ \mathbf{0} \end{bmatrix} \qquad (10.43)$$

to give

$$\mathbf{r}_s = -\mathbf{K}_{ss}^{-1}\mathbf{K}_{sm}\mathbf{r}_m \qquad (10.44)$$

and the transformation matrix is then

$$\mathbf{T} = \begin{bmatrix} \mathbf{I} \\ -\mathbf{K}_{ss}^{-1}\mathbf{K}_{sm} \end{bmatrix}. \qquad (10.45)$$

It is easy to show that the resulting condensed stiffness matrix preserves the static behaviour of the structure exactly. For a static analysis there is no approximation at all. However, there is an inherent approximation for dynamics because it has been assumed in equation (10.43) that the inertia forces at the slave freedoms are zero. This is only true in a very few cases. Some programs apply the Guyan reduction in a slightly different form, in that unit loads are applied to the master freedoms rather than unit displacements. The transformation matrix is then found by solving

$$\begin{bmatrix} \mathbf{K}_{mm} & \mathbf{K}_{ms} \\ \mathbf{K}_{sm} & \mathbf{K}_{ss} \end{bmatrix} \begin{bmatrix} \mathbf{r}_m \\ \mathbf{r}_s \end{bmatrix} = \begin{bmatrix} \mathbf{I} \\ \mathbf{0} \end{bmatrix} \qquad (10.46)$$

A FINITE ELEMENT PRIMER

giving

$$\mathbf{T} = \begin{bmatrix} \mathbf{I} \\ -\mathbf{K}_{ss}^{-1}\mathbf{K}_{sm} \end{bmatrix} (\mathbf{K}_{mm} - \mathbf{K}_{ms}\mathbf{K}_{ss}^{-1}\mathbf{K}_{sm})^{-1}. \quad (10.47)$$

Although, in practice **T** is found by solving equation (10.46) directly. It can be shown that this gives exactly the same eigenvalues for the condensed system as the Guyan reduction and the eigenvectors from the two techniques are related by the condensed stiffness matrix, as shown in equation (10.47). Hence the two methods are identical in principal and the choice of which one to use is then one of computational convenience.

The problem with all static condensation methods is the choice of the master freedoms. The aim is to choose these so that the resulting transformation matrix, **T**, is a linear combination of the first m eigenvectors. If **T** is deficient in any eigenvector then it will not be included in the analysis with consequential errors in the dynamic calculation. It will be impossible to choose the master freedoms so that **T** is an exact linear combination of the first m eigenvectors and there will always be a partial deficiency in the chosen subspace. This will have the effect of causing the eigenvalues of the condensed system to be higher than those of the full system. The master freedoms should be chosen so that they can define the low mode shapes and a first requirement for this is that they are distributed throughout the structure and not all, say, at the edges. One technique that is used for automatic selection of the masters involves taking the ratios of corresponding terms on the leading diagonals of the original stiffness and mass matrices and choosing the master freedoms to be those equations where this ratio is smallest. This has the effect of selecting freedoms that emphasise the high mass low stiffness parts of the structure and is in line with minimising the Rayleigh quotient and with equation (10.43) where the inertia forces at the slave freedoms were assumed to be zero. However, it fails in those cases where it concentrates all of the master freedoms in small regions. The use of this technique to automatically select master freedoms can be included in the matrix forward elimination (or factorisation) stage of the analysis. The first step of the subspace iteration method for finding eigenvalues and vectors (see section 10.2) corresponds to a Guyan reduction. (Subspace iteration itself can be considered as a repeated application of the Guyan reduction to make the transformation matrix for the condensation process the eigenvectors.)

A closely related condensation process is dynamic substructuring. For the Guyan reduction transformation of equation (10.45) the master freedoms are preserved identically as the freedoms of the reduced system. If the complete structure is considered to be a series of substructures and each substructure is condensed, using the freedoms that connect to other substructures as the master freedoms, then the resulting condensed component matrices can be assembled together using the standard finite element assembly process. This is the basis of dynamic substructuring. The simple method described here is not very accurate since the master freedoms are all along the edges that connect to other components and the resulting transformation matrix will not span the subspace defined by the lowest eigenvectors. The process can be improved by adding more vectors to the transformation matrix. There are various ways in which this can be done, the most common being to find the lowest mode shapes, ϕ, of the component with the master freedoms clamped. These are then added to $\mathbf{T}$ to give the new transformation as

$$\begin{bmatrix} \mathbf{r}_m \\ \mathbf{r}_s \end{bmatrix} = \begin{bmatrix} \mathbf{I} & 0 \\ -\mathbf{K}_{ss}^{-1}\mathbf{K}_{sm} & \phi \end{bmatrix} \begin{bmatrix} \mathbf{r}_m \\ \mathbf{q} \end{bmatrix}. \qquad (10.48)$$

The connection freedoms, $\mathbf{r}_m$, are still preserved and can be assembled whilst the other freedoms, $\mathbf{q}$, are internal to the component and do not get assembled to other components. This leads to the general requirement for the dynamic substructuring component transformation matrix of the form

$$\mathbf{T} = \begin{bmatrix} \mathbf{I} & 0 \\ \mathbf{A} & \mathbf{B} \end{bmatrix}. \qquad (10.49)$$

Provided $\mathbf{T}$ has this form the resulting condensed component matrices can be assembled. The various different dynamic substructuring methods then are all concerned with different methods for calculating $\mathbf{T}$.

Any dynamic subtructuring method is approximate since they all ignore some aspects of the mass distribution. As such they must be used with care so that the errors do not become significant. Unfortunately neither the various condensation or dynamic substructuring methods allow for estimates to be made of their accuracy relative to the solution that would be obtained from the full set of equations. About the only check that is available to the user is to use the Sturm sequence to verify that no modes have been missed over the range of interest.

There is one dynamic substructuring method that is exact provided that the response in the frequency domain is being found. This takes the dynamic stiffness matrix

$$\mathbf{K}(w) = (\mathbf{K} + iw\mathbf{C} - w^2\mathbf{M}) \qquad (10.50)$$

and substructures this when equation (10.24) is solved, precisely as for a static analysis. Since the dynamic flexibility matrix does not contain any functions of time then the substructuring is exact. However, it is not a numerically viable technique for anything other than structures with a small number of degrees of freedom since the substructuring has to be repeated from the beginning for each frequency that is solved.

10.8 Primary and Secondary Components

Most dynamic systems are not just simple structures but instead they consist of a complicated series of interconnected components. However, for a dynamic analysis it is usually only one or two of these components that are of direct interest and these can be termed the primary component(s). The analyst then has to decide in what detail the attached secondary components have to be modelled. To do this it is necessary to set up simple separate models of the individual primary and secondary components so that estimates can be made of their fundamental frequencies. The primary component is analysed with its natural boundary conditions and with any connections to secondary components left free. Each secondary component is analysed with points that connect to the primary held fixed, in addition to any natural support points. The fundamental (non zero) frequencies for the primary and secondary components can then be compared. If the fundamental frequency for the primary component is w_p and the fundamental for the secondary component currently under consideration is w_s then three possible situations can exist.

a. $0.5 \leq w_p/w_s \leq 2.0$,
b. $w_p/w_s < 0.5$,
c. $w_p/w_s > 2.0$.

If the frequency of the secondary component is close to that of the primary, as defined by (a) then they will interact dynamically and a dynamic model of the secondary component must be included together with the primary. This will be the case even if the mass of the secondary component is two orders of magnitude less than that of the primary. If condition (b) applies, that is the secondary component has a higher frequency than the primary, then the mass of the secondary can be added directly to the primary at the points where they connect. This is the

DYNAMIC ANALYSIS

natural assumption for handling the approximate modelling of the secondary component. However, if condition (c) applies, that is the secondary component has a lower frequency than the primary, then the stiffness of the secondary should be added to the primary and the secondary mass grounded. These various idealisations are shown in Figure 10.9.

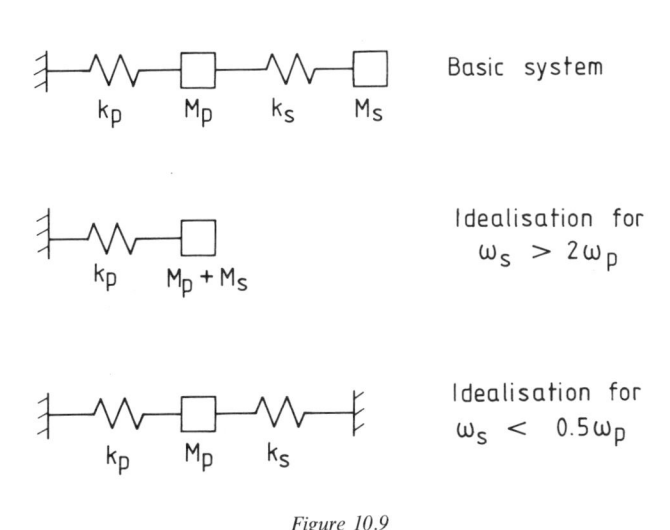

Basic system
k_p M_p k_s M_s

Idealisation for $\omega_s > 2\omega_p$
k_p $M_p + M_s$

Idealisation for $\omega_s < 0.5\omega_p$
k_p M_p k_s

Figure 10.9

Using the appropriate model for the secondary component the response of the primary can then be calculated. If the secondary component was modelled according to either conditions (b) or (c) then its dynamic response can be found by taking the movements computed from the primary response and applying these as base movements at the points of connection between the primary and secondary components to the separate secondary component model. This is illustrated in Figure 10.10.

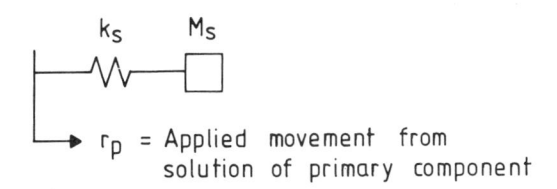

k_s M_s

r_p = Applied movement from solution of primary component

Figure 10.10

201

These rules of thumb for modelling primary and secondary components are all based upon the approximations that can be made to a two degree of freedom system. It is not possible to define precise rules for a multi-degree of freedom system but the assumption used here is that the fundamental response of each component is dominant and the higher frequencies can be ignored for modelling purposes, especially since the response of the secondary component is not actually required.

10.9 Number of Modes for a Dynamic Analysis

So far this discussion of modal solutions parallels other texts in that it is assumed that only the first few modes are used for the analysis, without being precise as to what is meant by the number of modes that are actually required. The only dynamic solution technique that commonly considers this problem is seismic analysis using the response spectrum method. Whilst this technique is perfectly valid for that problem, the conclusion that it leads to for the typical number of modes required is an extreme result which arises from the nearly uniform distribution of loads over the structure in the seismic case. It is necessary to generalise these ideas to other forms of loadings for more general conclusions and this is done in this section.

The response of any structure to a general time history is given by the convolution (or Duhamel) integral of equation (10.19). The force input can be rewritten as a summation involving the various time histories of the forcing functions

$$\mathbf{R}(t) = \sum_i a_i(t)\mathbf{R}_i. \qquad (10.51)$$

Very often there is only one time history involved so that an excitation of the form

$$\mathbf{R}(t) = a(t)\mathbf{R} \qquad (10.52)$$

can be considered. Here **R** defines the spatial distribution of the force over the structure and $a(t)$ is the scalar time history. If the modes required to find the response to this force input can be defined then repeated application to each term in the series of equation (10.51) allows the number of modes for any force input to be found. This requires that some form of convergence test is defined. The measure used here is to take the scalar product of the force distribution multiplied by the absolute

DYNAMIC ANALYSIS

maximum displacement, giving a work quantity

$$u = \mathbf{R}^t \max |\mathbf{r}(t)|. \quad (10.53)$$

Substituting for the displacement at any time, using equation (10.19), and manipulating the equations gives this measure as

$$u = \sum_{i=1}^{m} \mathbf{Q}_i^t \mathbf{Q}_i \max \left| \int_0^t \frac{e^{-\alpha_i(t-\tau)} \sin \beta_i(t-\tau) a(\tau) \, d\tau}{m_i \beta_i} \right| = \sum A_i B_i. \quad (10.54)$$

Note that here the response has been separated into the time varying part, B_i, and the spatially varying part, A_i. Both of these depend upon the number of modes that are taken in the analysis and their convergence can be investigated independently.

For the spatial variation then the convergence of

$$A = \sum_{i=1}^{m} \mathbf{Q}_i^t \mathbf{Q}_i \quad (10.55)$$

is required, where $\mathbf{Q}_i$ is the modal (or generalised) force for the ith mode. If all of the modes are taken in the summation and they are normalised such that

$$\boldsymbol{\phi}^t \mathbf{M} \boldsymbol{\phi} = \mathbf{I} \quad (10.56)$$

then it can be shown that

$$\sum_{i=1}^{n} \mathbf{Q}_i^t \mathbf{Q}_i = \mathbf{R}^t \mathbf{M}^{-1} \mathbf{R} \quad (10.57)$$

and this can be found without knowing any of the systems eigenvectors. Thus as more modes, m, are taken then the spatial variation tend to the limit

$$\sum_{i=1}^{m} \mathbf{Q}_i^t \mathbf{Q}_i \to \mathbf{R}^t \mathbf{M}^{-1} \mathbf{R} \quad (10.58)$$

and the solution can be considered to have converged in the spatial sense if m is such that

$$\sum_{i=1}^{m} \mathbf{Q}_i^t \mathbf{Q}_i \geq f \mathbf{R}^t \mathbf{M} \mathbf{R} \quad (10.59)$$

A FINITE ELEMENT PRIMER

where f is some number close to unity, typically 0.9. This is illustrated by the example of the simple framework shown in Figure 10.11.

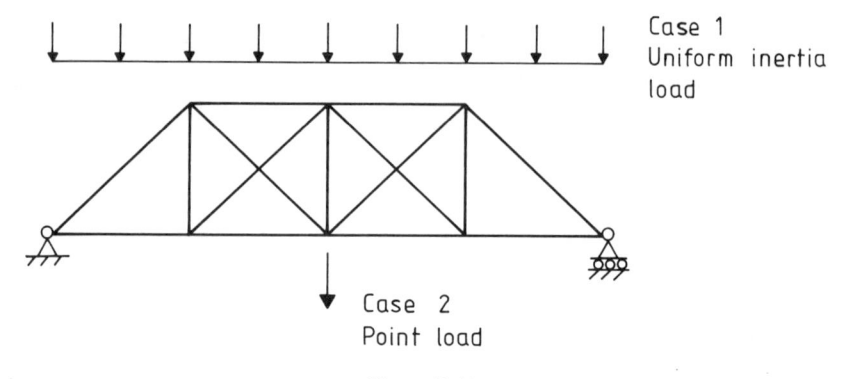

Figure 10.11

The convergence for the spatial terms for a uniformly distributed acceleration loading (a seismic loading) and the point load case is given in the following table.

Mode no.	Mode participation factor Case 1	Case 2
1	0.699	0.159
2	0.004	0.000
3	0.010	0.149
4	0.010	0.084
5	0.000	0.028
6	0.002	0.172
7	0.005	0.211
8	0.000	0.114
9	0.100	0.052
10	0.171	0.030

It will be seen from this that the distributed inertia load case converges very quickly with most of the response occurring in the first mode, a typical result for seismic analysis. However, the point load case requires all of the modes to be present before the spatial variation has converged and rather more modes are always required for discrete loadings compared to distributed loads.

DYNAMIC ANALYSIS

The time variation can be considered by investigating how the function

$$B = \max \left| \int_0^t e^{-\alpha(t-\tau)} \sin \beta(t-\tau) \frac{a(\tau)}{\beta} d\tau \right| \qquad (10.60)$$

varies as the damped frequency, β, is changed. The function will eventually die away to zero as the frequency is increased, since a finite value at a high frequency implies that the system has a very large amount of energy. The time history $a(t)$ can be characterised by evaluating equation (10.60) over a range of frequencies and plotting B against frequency. This gives the response spectrum curve for $a(t)$ and it can be found for typical values of damping before any of the structures eigenvalues are evaluated. It is then used to define the highest natural frequency that has to be extracted for the system. Typically if the response spectrum peaks at a value of B_m then a cut off frequency, w_c, can be defined, such that $B < f B_m$ for all frequencies above w_c, where f is some small number, typically 0.1. This is then used to define the highest resonant frequency that need be found in the eigenvalue extraction.

The combination of the spatial and the time terms give the convergence requirements for the displacement response. The response spectrum is closely related to the Fourier transform of the time history and often this can be used instead of the response spectrum curve itself. The process can be repeated for the velocities. The spatial variation convergence is as for the displacements. However, the time variation, as defined by the response spectrum, for the velocity becomes approximately

$$\dot{B} = \max \left| \int_0^t e^{-\alpha(t-\tau)} \sin \beta(t-\tau) a(\tau) d\tau \right|. \qquad (10.61)$$

This assumes that the system is lightly damped. The convergence of the velocity response spectrum will be slower than that for the displacements implying that more modes are required for a velocity response than for the displacement. Similarly the acceleration response spectrum is

$$\ddot{B} = \max \left| \int_0^t \beta e^{-\alpha(t-\tau)} \sin \beta(t-\tau) a(\tau) d\tau \right| \qquad (10.62)$$

and this will converge even more slowly than the velocity. In fact there are some time histories where the acceleration response spectrum does not tend to zero at high frequencies and the convergence for accelerations is controlled entirely by the convergence of the spatial variation of the force.

10.10 Calculation of Dynamic Stresses

The previous section has defined how a measure can be established for the convergence of the displacements, velocities and accelerations. It can be shown for beam type structures that the stresses converge at a rate very similar to that for the velocities. This is probably also true for the overall average stress levels for any structure. However, for structures that have any form of stress concentration it is not the average stress levels that are of interest but the peak values at the stress concentrations. These involve small local regions of high strain energy with low relative velocities around the stress concentration. It follows from consideration of the Rayleigh quotient that such a behaviour will be associated with high frequency modes and the low frequency ones will not excite the stress concentrations very strongly. This argument implies that if any form of condensation is used to calculated the dynamic response then this will filter out the peak stresses and they will be considerably underestimated. Such a situation will arise if the stresses are determined directly from the displacement response. The stresses are underestimated no matter how many degrees of freedom were used in the original model since the error is directly related to the degree of condensation used, not upon the refinement of the original mesh. The error arises for any form of condensation, whether it is by means of a reduced number of modes, Guyan reduction, dynamic substructuring or even unconditionally stable step by step integration methods with relatively large timesteps in comparison to the mid range frequencies.

A much better estimate of the peak stresses can be found from the response, provided that the user is prepared to use more computer time. The accelerations that are found at any stage of the response calculation will be reasonably accurate provided that the condensation used was not too severe. These accelerations can then be used at any time to define the inertia forces on the structure as

$$\mathbf{R}_I = -\mathbf{MT\ddot{q}} \qquad (10.63)$$

where $\mathbf{T}$ is either the eigenvectors of equation (10.12), in which case $\ddot{\mathbf{q}}$ is the modal acceleration response, or $\mathbf{T}$ is the condensation matrix of equation (10.43), in which case $\ddot{\mathbf{q}}$ is the master freedom acceleration vector $\ddot{\mathbf{r}}_m$. These can then be applied as static forces and the stresses found by solving the static problem

$$\mathbf{Kr}(t) = \mathbf{R}(t) - \mathbf{MT\ddot{q}}(t). \qquad (10.64)$$

This uses the full mesh for the model and the accuracy of the stress distribution will then be comparable to a static analysis with the same mesh. The method works because the high frequency modes that define the stress concentration have a constant response over the low frequency range that has been considered for the dynamic calculation and, since the stresses are found from an instantaneous static sum, the constant components of the high frequency response have been included in equation (10.64). Also, the relative accelerations around the stress concentration are low and all that is required is a reasonably accurate estimate of the overall structural accelerations. This will be given by the lower modes. The user can test the accuracy of the stress recovery process of any system by applying a constant (that is static) load to the dynamic analysis and calculating the corresponding stress distribution produced when the dynamic response has become stationary. This can then be compared with the stress distribution produced directly by a true static analysis. If the two stress states are comparable then the dynamic stress recovery can be considered verified for that geometry and that distribution of loading.

10.11 Result Recovery

A problem with a dynamic analysis that does not occur with statics is the sheer volume of data that can be produced. If the displacements and stresses are found at every time step for a time domain solution then, assuming 20 steps per cycle of the lowest mode and assuming only 100 cycles of the response, the equivalent of 2000 static load cases will be printed for this one single time solution. Obviously this is a very large volume of information and only a very small fraction of it is of interest. The user must consider how to control this data before he embarks upon the solution, ensuring that the important values in the response are displayed without being swamped in output. For a modal solution this can be done by carrying out the analysis in stages. The first step involves finding the eigenvalues and vectors of the structure and calculating the modal stresses by treating the eigenvectors as displacements. The magnitude of these stresses will have no meaning but their distribution will indicate the regions of high stress so that the user can arrange to only recover stresses in these areas. The second step is to find the time history responses of the individual modal equations so that the times at which the individual modal displacements peak can be found. The third step is to then recover the stresses in the regions of high stress for the periods of time around the time when the modal responses peak. This will ensure that the peak dynamic values are found but only the minimum amount of information is handled. It is not possible to say anything about the relative magnitudes of the modal responses since the physical meaning of these will depend upon the size of the terms in the eigenvectors, which in turn depend upon the normalisation that is used. The modal responses will probably be related to the internal eigenvector normalisation, which might well be different from that used for the display (printing) of the vectors, thereby complicating further the interpretation of the modal stresses.

10.12 Structural Modifications and the Dynamic Analysis

Since a dynamic analysis is relatively expensive it is necessary to have available approximate techniques for evaluating the effect of any changes to the structures geometry. This might arise either because of changes to the structure after the analysis has been conducted or the need to alter it to modify the dynamic response itself. If the eigenvalue problem

$$\mathbf{K}\boldsymbol{\phi}_i = \lambda_i \mathbf{M}\boldsymbol{\phi}_i \tag{10.65}$$

has been solved so that the eigenvalues, λ_i, and the eigenvectors, $\boldsymbol{\phi}_i$, are known then the effect of adding extra small stiffness terms, $\mathbf{K}_\Delta$, and small mass terms, $\mathbf{M}_\Delta$, to the equations have to be found. This could be done by solving the new eigenvalue problem

$$(\mathbf{K}+\mathbf{K}_\Delta)(\boldsymbol{\phi}_i+\boldsymbol{\phi}_{i\Delta}) = (\lambda_i+\lambda_{i\Delta})(\mathbf{M}+\mathbf{M}_\Delta)(\boldsymbol{\phi}_i+\boldsymbol{\phi}_{i\Delta}) \tag{10.66}$$

but this is expensive. Instead a first order expansion of equation (10.66) can be used to obtain the approximate variations in the eigenvalues and vectors. The first order change in the eigenvalue is

$$\lambda_{i\Delta} = \frac{\boldsymbol{\phi}_i^t(\mathbf{K}_\Delta - \lambda_i \mathbf{M}_\Delta)\boldsymbol{\phi}_i}{\boldsymbol{\phi}_i^t \mathbf{M}\boldsymbol{\phi}_i} \tag{10.67}$$

and the corresponding first order change in the eigenvector is

$$\boldsymbol{\phi}_{i\Delta} = \sum_{\substack{j=1 \\ j \neq i}}^{m} \boldsymbol{\phi}_j \alpha_j \tag{10.68}$$

where

$$\alpha_j = \frac{\boldsymbol{\phi}_j^t(\mathbf{K}_\Delta - \lambda_i \mathbf{M}_\Delta)\boldsymbol{\phi}_i}{(\lambda_i - \lambda_j)\boldsymbol{\phi}_j^t \mathbf{M}\boldsymbol{\phi}_j}. \tag{10.69}$$

DYNAMIC ANALYSIS

If ϕ only contains the first few modes then this estimate for the modification to the vector will only be very approximate but the variation to the eigenvalue, $\lambda_{i\Delta}$, has been found to be surprisingly accurate, even for relatively large changes to the mass or stiffness.

10.13 Wave Propagation

A special dynamic problem is the analysis of the propagation of a wavefront through a medium. This can be solved using the standard analysis techniques described in section 10.5 provided that various precautions are taken. As the wave propagates into the model then portions of the structure that are initially at rest are suddenly forced into movement as the wavefront passes. The highest frequency of any wave that can be propagated then depends upon distance between consecutive nodes. If the highest frequency to be propagated is f Hz and the wave velocity is c m/sec then the wavelength is

$$l = c/f. \tag{10.70}$$

There should be at least four nodes over this wavelength in order for the wave to propagate. A uniform mesh of elements that have all sides of the same length should be used to allow waves to propagate equally in any direction. If a change in mesh density is used within the model then this gives rise to a change in the effective numerical impedance where the mesh density change occurs. Spurious internal wave reflections can occur at such changes so that the correct propagation behaviour is not reproduced. This type of wave propagation is ideally suited to solution by conditionally stable step-by-step integration methods where the time step used should be slightly less than the time that it takes the wavefront to travel between two adjacent nodes. If the time step is any longer than this there is an effective truncation in the highest frequency that can be propagated, and if it is shorter then an excessive amount of computation will be carried out. Although the high frequency modes of the equations of motion have no physical meaning they are important numerically for wave propagation problems. If they are not included then the wavefront gets dispersed and ceases to be as sharp as it should be. Solution methods that include all of the information contained in the original equations of motion will give better answers than those that truncate the data. These include modal condensation, static condensation (Guyan reduction), sub structuring and unconditionally stable step-by-step methods where the time step is significantly greater than the time that it takes to propagate the wave between two adjacent nodes. Under these conditions it is almost invariably cheaper to use a conditionally stable integration method where the requirements for the choice of the timestep will guarantee convergence. The higher the degree of condensation that is used (the more high frequency modes that are dropped) then the greater will be the dispersion (or smearing) of the wavefront.

10.14 Seismic Analysis

There is much interest currently in proving that large structures and associated components can resist the effects of an earthquake. The main difficulty with a seismic analysis compared to any other form of dynamic analysis is that the forcing function is a non-stationary random quantity. Methods for solving such types of input are not well developed and some form of approximate solution is usually employed. This can take one of two forms. Firstly, a series of earthquake time histories that are typical of the location of the plant are defined. The response to all of these different inputs are then found using standard time domain solution methods. The peak values of a given response quantity, say the stress at a point, is then found for all times and all of the histories and this is used as the seismic response. In the second method the response spectrum, as defined in section 10.9, is found for the various earthquakes. These are plotted on the same axes, a smooth envelope is drawn around them and this is taken to define the working response spectrum for the analysis.

Two levels of seismic event are defined for a given geological region, the safe shutdown earthquake and the operating basis earthquake. For safe shutdown, as the name implies, it must be demonstrated that the plant can be safely switched off after the seismic event even though it might not function again. This means that the input was sufficiently strong to cause permanent damage which implies that the response is non-linear. With current solution techniques this requires that the response is found by a step-by-step numerical integration of the non linear equations for a series of input histories. The analyst then has to demonstrate that the plant can be shut down after it has been subjected to any one of the postulated earthquakes. For the operating basis event it is necessary to show that the plant is still functioning after the excitation. This implies that the response is linear throughout the event and various solution techniques can then be used. The time history response to all of the design earthquakes can be found from either a modal or a step-by-step solution. This is usually done in those cases where the response of secondary components attached to the plant has to be found since the input to these is then available from the primary response. Alternatively the response spectrum method can be used where appropriate to give a very much simpler response calculation. However, there are various problems associated with this method, all arising from the fact that only the peak response was considered in calculating the response spectrum in the first place. This means that there is no mathematically rigorous method for recombining the individual modal responses. If the natural frequencies are well separated the combination is usually done on the basis of the square root of the sum of the squares (SRSS) of the modal responses. For those modes which have close natural frequencies then an absolute sum is taken on the assumption that they can peak in phase with each other. Alternatively a complete quadratic combination (CQC) method can be used which effectively combines both methods so that it degenerates to either depending upon how close the eigenvalues are to each other. A variety of other modal combination methods have also been defined for combining various directions of excitation. The second restriction associated with the response spectrum method is that it is only simple if all of the support movements in a given direction are identical. This is usually valid for ground movements applied to buildings or to local floor movements applied to equipment but it is not applicable to piping systems where the anchor points of the pipe are connected to different structures or even different parts of the same structure. When using the response spectrum method the results cannot be processed other than to find the maximum values. All of the displacements and stresses will always be positive since they are combined by an absolute sum or an SRSS method.

When the ground movements in a given direction are all identical then the structural response can be found either absolutely or relative to the ground movement. The stresses will be the same in both cases. Usually a relative calculation is conducted but some programs work in terms of absolute values. For an absolute response then the user must ensure that the input earthquake time history is baseline corrected, that is, when it is integrated the resulting velocity and displacement are both zero. The usual form of the equation of motion for the relative response is

$$\mathbf{M}\ddot{\mathbf{r}} + \mathbf{C}\dot{\mathbf{r}} + \mathbf{K}\mathbf{r} = -\mathbf{M}\mathbf{e}a(t) \qquad (10.71)$$

where $\mathbf{r}$ is the displacement relative to the ground, $a(t)$ the applied ground acceleration and $\mathbf{e}$ the vector defining the direction of the ground movement. A rigorous development of this equation shows that it implies that there is no mass coupling between the structure and the ground (the

A FINITE ELEMENT PRIMER

off-diagonal terms in the unsupported mass matrix that couple the structure and the ground movements are zero or (are insignificant) and that there is no damping to ground. These are usually good approximations but in those cases where they are not correct then an absolute response form should be used.

10.15 Random Vibrations

The forcing functions for a dynamic analysis are sometimes random in nature, that is a precise value can not be defined at any time. Only average values in terms of mean, mean square and higher order moments can be defined

mean $\quad\quad\quad\quad\quad\quad \mu = E(x) \quad\quad\quad\quad$ (10.72)

mean square $\quad\quad\quad\quad \sigma = E[x^2]. \quad\quad\quad\quad$ (10.73)

The correlation can also be defined, where this is the average of the product of the response with a constant time delay, τ, between the two samples,

correlation $\quad\quad R_{xx}(\tau) = E[x(t)x(t+\tau)]. \quad\quad$ (10.74)

If these and similarly defined average quantities remain constant with time then the process is said to be stationary. If they vary with time the process is said to be non-stationary. There are no general techniques for the analysis of non-stationary inputs and the approximate methods for the seismic analysis of section 10.14 can be used. For a stationary random input it is usually necessary to calculate the corresponding stationary random response since any initial start up transients are not significant. As for the calculation of the deterministic periodic response, this is best done in the frequency domain. Here the spectral density of the force, $S_{RR}(w)$, is the Fourier transform of the correlation given by equation (10.74) and it is defined for a range of frequencies, w. In general this will be a complex matrix, with a symmetric real part and a skew symmetric imaginary part. The spectral density of the displacement response can then be easily calculated from

$$S_{rr}(w) = \bar{H}(w) S_{RR}(w) H^t(w) \quad\quad (10.75)$$

where $H(w)$ is the structure's dynamic flexibility matrix as defined by either equation (10.24) or (10.25). $\bar{H}(w)$ is the complex conjugate of the dynamic flexibility. The other response spectral densities that can be computed from this are:

$$S_{r\dot{r}}(w) = iw S_{rr}(w), \tag{10.76a}$$

$$S_{\dot{r}r}(w) = -iw S_{rr}(w), \tag{10.76b}$$

$$S_{\dot{r}\dot{r}}(w) = w^2 S_{rr}(w), \tag{10.76c}$$

$$S_{r\ddot{r}}(w) = -w^2 S_{rr}(w), \tag{10.76d}$$

$$S_{\dot{r}\ddot{r}}(w) = iw^3 S_{rr}(w), \tag{10.76e}$$

$$S_{\ddot{r}r}(w) = -w^2 S_{rr}(w), \tag{10.76f}$$

$$S_{\ddot{r}\dot{r}}(w) = -iw^3 S_{rr}(w), \tag{10.76g}$$

$$S_{\ddot{r}\ddot{r}}(w) = w^4 S_{rr}(w). \tag{10.76h}$$

The mean square of the responses can be found by integration

$$E[\mathbf{r}^2] = 2 \int_0^\infty S_{rr}(w)\,dw, \tag{10.77a}$$

$$E[\dot{\mathbf{r}}^2] = 2 \int_0^\infty S_{\dot{r}\dot{r}}(w)\,dw, \tag{10.77b}$$

$$E[\ddot{\mathbf{r}}^2] = 2 \int_0^\infty S_{\ddot{r}\ddot{r}}(w)\,dw. \tag{10.77c}$$

When carrying out these integrations it must be remembered that significant peaks will occur at the various structural resonant frequencies and most of the response will be at the resonances. This means that the resonant peaks should be well defined for equation (10.77) but there is no need for such precision away from resonance. The spectral density and mean square values of the stresses can be found for the stationary random response in a similar manner.

10.16 Some useful references

● W.C. HURTY and M.F. RUBENSTEIN, Dynamics of Structures, Prentice Hall, 1974.

A FINITE ELEMENT PRIMER

- J.H. WILKINSON, The Algebraic Eigenvalue Problem, Clarendon Press, Oxford, 1965.
- K.J. BATHE and E.L. WILSON, Numerical Methods in Finite Element Analysis, Prentice Hall, 1976.
- R.R. CRAIG, Structural Dynamics, Wiley, 1981.
- R.W. CLOUGH and J. PENZIEN, Dynamics of Structures, McGraw-Hill, 1975.
- Y.K. LIN, Probalistic Theory of Structural Dynamics, McGraw-Hill, 1967.

11. Nonlinear Analysis

11.1 Introduction

All the analysis dealt with so far has been linear, and this chapter deals with the use of the finite element method to analyse nonlinear problems. By waiting until chapter 11 the implication seems to be that nonlinear finite element analysis is either 'difficult' or 'unimportant', and both of these views have been stated in the past by respectable engineers. Indeed if we look at the proportion of finite element analysis which is analysed in a true nonlinear fashion, that proportion is small. In the 1970s it was probably less than 1%, in the early eighties it had risen to between 5% and 10% and the proportion is still growing as the cost and accessibility of computers improves. Nonlinear analysis was in effect an academic pursuit (and hence warranted the description 'difficult' perhaps) until reliable finite element software became available, which is the reason for this chapter.

The other adjective 'unimportant' is of course subjective but does have some validity when we look at the way in which practising engineers choose, or are forced, to use the stresses and strains which their analysis delivers. Invariably some failure criteria has to be checked. The simplest one is an unacceptable deformation clearly, but this is rarely the overriding failure criteria in any structural field. The maximum load which a structure can carry safely is usually the ultimate condition, and this is nearly always related to a stress level. A ductile metal might deform plastically in such a way that the structure becomes a mechanism with no stiffness. A structure or component part may buckle and again lose its stiffness. A less ductile material may attain a stress at which a likely flaw develops into an unstable crack under constant load. Alternatively if the loading is unsteady the material may suffer a fatigue failure or have cracks grow until they become of unstable size again. Other forms of material instability are also possible like creep-rupture, particularly at high temperatures.

There are many other forms of failure, but the one thing that all these failure modes have in common is that *they are all nonlinear* if an attempt is made to analyse them in detail. Either the stress is no longer proportional to strain or else there are such gross changes in geometry that they cannot be ignored. And yet engineers have almost never used nonlinear analysis as a routine failure diagnostic. Firstly because the finite element method had not yet arrived, secondly because the physical modelling was not well understood, thirdly because a numercial analogue was extremely expensive, and fourthly because the local failure criteria was available anyway in the forms of experimental results, quasi-empirical data sheets, codes of practice and so on. The fourth alternatives are still the most popular for local failure criteria, but the first three objections are gradually becoming less credible, as we mentioned. We therefore turn to nonlinear analysis, especially using the finite element method.

The approach used so far in this primer assumes a linear relationship at several stages. Firstly we assume that the strains are small, and so dimensional changes in materials thickness, or cross-sectional area for example, are negligible. This assumption will continue to be made since most materials have outlived their usefulness when the strain exceeds one or two per cent. The exceptions are plastics and rubbers, and then a large strain analysis may be necessary. This is such a specialised field that we ignore it.

Secondly the compatibility relationships, $\varepsilon = \partial \mathbf{u}$, may cease to be linear if the displacements are large even though the strains are still small. The differential operators in ∂ will no longer be linear. The changes in the geometry of the deformed shape can no longer be ignored. Buckling is one such phenomenon where, if geometry changes are ignored, then no buckling load can be predicted. Some structures like cables or membranes are excessively flexible until the loads induce respectable stresses and displacements. Yet others have component parts which may separate or make contact as the load is changing.

Thirdly the stress-strain law may cease to be linear, even within the useful stress range of the material. This inelastic behaviour may also occur simultaneously with buckling.

Fourthly the geometrical changes in the shape of the structure may be so significant that the original equilibrium equations, relating stress to loads, will have to be updated. Fortunately the finite-element displacement method uses the PVD to satisfy equilibrium and we only have to ensure that the presence of large displacements does not invalidate it.

NONLINEAR ANALYSIS

Conventional structural analysis consequently has three linear links all of which may be broken.

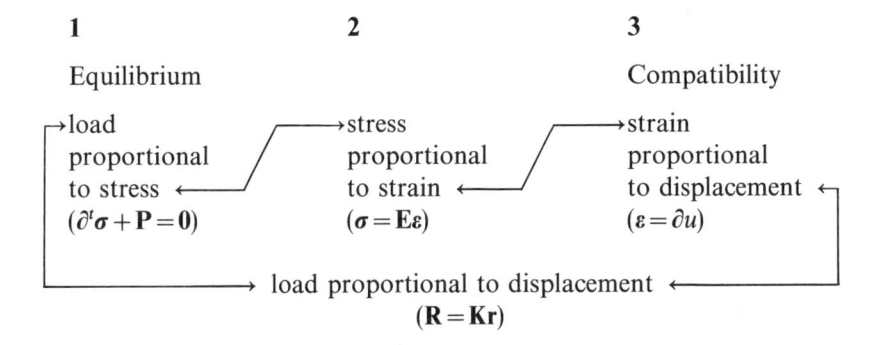

```
    1                       2                         3
Equilibrium                                      Compatibility
 ┌→load              ┌→stress              ┌→strain
 │ proportional     /  proportional       /  proportional
 │ to stress    ←──/   to strain    ←────/   to displacement ←┐
 │ (∂'σ + P = 0)       (σ = Eε)              (ε = ∂u)         │
 │                                                            │
 └──────────→ load proportional to displacement ←─────────────┘
                        (R = Kr)
```

The nonlinear inelastic behaviour is dealt with in 11.4, and we turn first to the problem of gross deformations ($\varepsilon \neq \partial \mathbf{u}$).

11.2 Gross Deformations

A very simple example serves to illustrate many of the problems (and solutions) present during gross deformations.

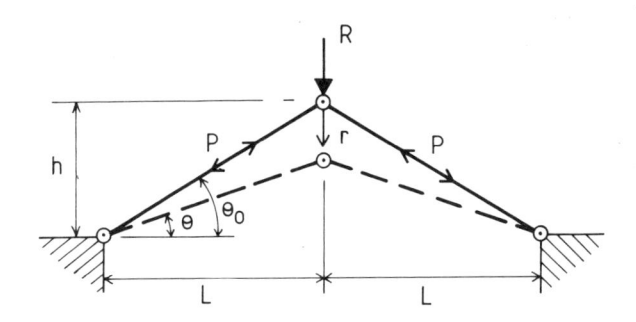

Figure 11.1 A shallow framework.

The two bar framework shown in Figure 11.1 has a rise, $h = l \tan \theta_0$, which may be so shallow that even a small joint displacement r causes a modification to the original geometry: that is r is a functions of load. However in this simple model it is possible to write down the three fundamental equations for both small r or large r, without any

A FINITE ELEMENT PRIMER

approximations

Equilibrium: $$R = 2P \sin \theta \quad (11.1)$$

Compatibility: The compressive strain

$$\varepsilon = \left(\frac{l}{\cos \theta_0} - \frac{l}{\cos \theta} \right) \bigg/ \frac{l}{\cos \theta_0} = 1 - \frac{\cos \theta_0}{\cos \theta} \quad (11.2)$$

where $\tan \theta = (h - r)/l$.

We assume moderately small strains in the bars (of cross-sectional area A) but not so excessive as to invalidate Hooke's Law:

$$\varepsilon = P/AE. \quad (11.3)$$

The load/displacement relationship is shown in Figure 11.2 for two values of the initial position, $\theta_0 = 10°$ and $\theta_0 = 30°$.

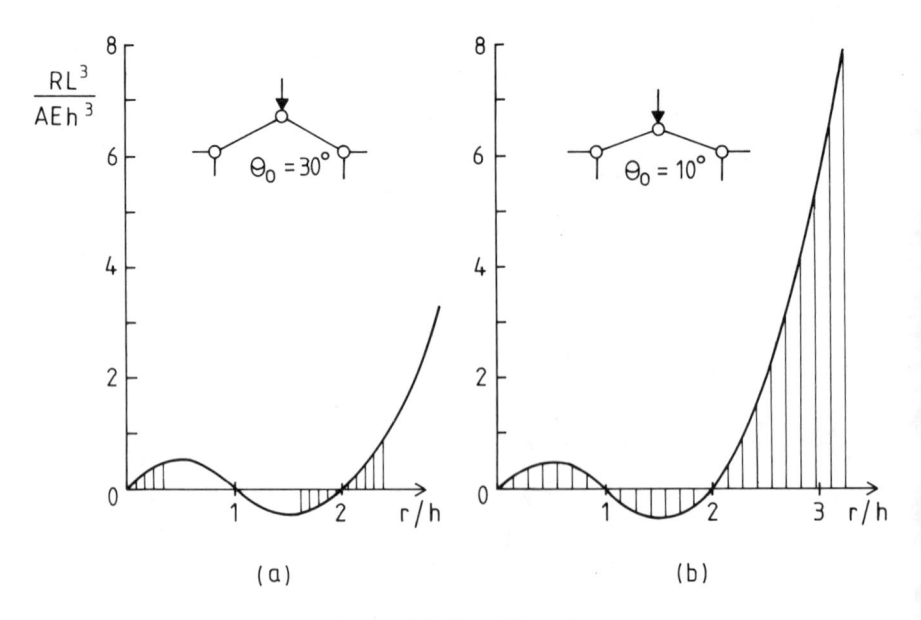

Figure 11.2 Snap Through.

NONLINEAR ANALYSIS

On these two graphs are marked the region where the strain is less than 5% to justify the use of Hooke's Law. In fact very few materials remain linear as far as this. The difference between the two cases is clear. In the non-shallow framework (Figure 11.2a) there are two regions where the strains are small, one for $r/h<0.2$ and the other near $r/h=2\pm0.2$ where the arch has 'snapped through' to its mirror image $\theta=-\theta_0$. In both regions the load-displacement relationship is approximately linear, and changes in the geometry from the unloaded position can be ignored when writing down equilibrium conditions. So (11.1) becomes

$$R = 2P \sin \theta_0.$$

The equations of compatibility (11.2) can be approximated as

$$\varepsilon = \sin \theta_0 \cos \theta_0 \cdot r/l$$

and we find the expected linear relationship

$$R = \frac{2AE \sin^2 \theta_0 \cos \theta_0}{l} \cdot r. \tag{11.4}$$

The solution in the case of the shallow arch (Figure 11.2b) is quite different. Even though the strains are small, the force-displacement relationship is decidely non-linear, and changes in geometry cannot be ignored. It is possible to solve this problem once we recognise that h/l, θ, and θ_0 are all small and we can approximate the equilibrium equation (11.1) as

$$R = 2P(h-r)/l. \tag{11.5}$$

Compatibility (11.2) becomes

$$\varepsilon = 1 - \left(\frac{h}{l}\right)^2 \frac{r}{l} + \tfrac{1}{2}\left(\frac{r}{l}\right)^2 + \cdots \tag{11.6}$$

On substituting $P = AE\varepsilon$ into (11.5) and using (11.6) we find

$$\frac{Rl^3}{AEh^3} = \left(\frac{r}{h}\right)\left(1-\frac{r}{h}\right)\left(2-\frac{r}{h}\right)$$

which is a nonlinear cubic as suggested by Figure 11.2b.

219

A FINITE ELEMENT PRIMER

But this is a very simple problem, so how can the finite element method be adopted to solving such problems with many degrees of freedom? More importantly how can nonlinear compatibility equations of the complexity of (11.2) be derived? The trick is to consider *small* incremental displacements $\delta \mathbf{r}$ due to small load increments $\delta \mathbf{R}$ away from the grossly deflected position. If the incremental displacements $\delta \mathbf{r}$ are small, due to small loads $\delta \mathbf{R}$, the problem is linear in $\delta \mathbf{r}$ and $\delta \mathbf{R}$ except that the total displacements $\mathbf{r}$ and the current stresses have to be summed over all previous results—that is the history of the structure has to be constructed. In the problem of our shallow arch, we consider a small load increment from R to $R+\delta R$ where (11.1) tells us that

$$\delta R = 2\delta P \sin \theta + 2P \cos \theta \cdot \delta \theta \qquad (11.7)$$

and *the change in geometry $\delta \theta$ is not ignored*. Because δP is a small increase in stress, we can use the linear compatibility relationship between strain and displacement using the current geometry θ,

$$\delta \varepsilon = \sin \theta \cos \theta \cdot r/l$$

and

$$\delta P = AE \sin \theta \cos \theta \cdot \delta r/l.$$

Having $\tan \theta = (h-r)/l$, we find $\delta \theta = \cos^2 \theta \cdot \delta r/l$ so (11.7) becomes

$$\delta R = \left(\frac{2AE \sin^2 \theta \cos \theta}{l} - \frac{2P \cos^3 \theta}{l} \right) \delta r$$

$$= k_T \delta r.$$

This relationship between incremental load δR and displacement δr is called a *tangent stiffness* k_T (see Figure 11.3). It has two components, firstly the *elastic stiffness*

$$k_E = 2AE \sin^2 \theta \cos \theta / l$$

which is the usual stiffness (11.4) except that θ is now the current changing value of θ and may differ significantly from the original value θ_0. But the second term is additional and represents the resistance to load caused by realigning the internal stresses (P) when displacements occur. This stiffness is called the *geometric stiffness* k_G and is evaluated in terms of the *current stresses*—in this case the single constant P.

NONLINEAR ANALYSIS

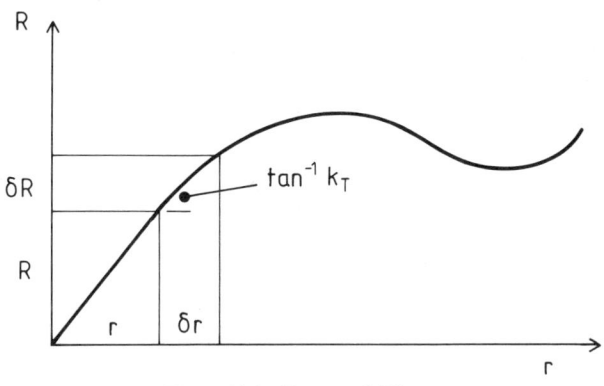

Figure 11.3 Tangent Stiffness.

Thus

$$k_G = \frac{2P\cos^3\theta}{l}$$

and

$$k_T = k_E - Pk_G. \tag{11.8}$$

In solving this sort of problem numerically the load could be increased in small increments, and the value of k_T found in terms of θ and P which have to be accumulated from previous increments. After each increment the displacement δr is found from $\delta r = k_T^{-1} R$, and the next value of k_T found, and so on.

For more general structures the geometric stiffness and the total tangent stiffness for each element has to be constructed and assembled into a global tangent stiffness in the familiar manner, but recognising that the positions of every element may now change as the loads increase. One approach is to refer everything to a fixed set of axes, and use expressions like (11.2) in terms of the total displacements (the total Lagrangian method). This can be complicated and is rarely done. Another (Eulerian) approach is to carry a local set of axes with every deformed element in the structure, no matter how gross the deformations. The equations governing the behaviour are thereby modified but at least the displacements are small with respect to the convected axes. This approach is also rarely used in structures, but it is common in aerodynamics/fluid mechanics where the absolute movements of deformed material may be vast.

The most common approach is a compromise (the updated Lagrangian) whereby the local axes are updated but then the element displacements are assumed to be *moderate* with respect to these axes. 'Moderate' means that rotations – in radians – when squared, are small compared to one. So five degrees is small. The derived element stiffness matrix can always be transformed to fixed global axes using the transformation (4.3). By allowing only moderately large displacements from the updated local axes, it is possible to construct geometric stiffness matrices in a methodical fashion. For example suppose we wished to evaluate the geometric stiffness of the beam element, for which we previously derived the elastic stiffness (4.8), by using the Engineers' theory of bending again: (4.1)

$$\dot{w} = w_0 - yv'(z). \qquad (4.1)$$

It is readily shown that the general form of (11.6) for moderately large displacements $w(z)$ and $v(z)$ is

$$\varepsilon = \frac{\partial w}{\partial z} + \tfrac{1}{2}\left(\frac{\partial v}{\partial z}\right)^2$$

or

$$= w' + \tfrac{1}{2}(v')^2$$

for brevity.

So substituting (4.1) the nonlinear strain-displacement relationship for a beam becomes

$$\varepsilon_{zz} = w'_0 - yv'' + \tfrac{1}{2}(v')^2.$$

Therefore small displacements $\delta v(z)$ will produce increments

$$\delta \varepsilon_{zz} = -y\delta v'' + v'\delta v'.$$

The incremental expression demonstrates that small strain increments $\delta \varepsilon$ are indeed proportional to small displacements δv, but they also contain the total current displacement v – which has a nonlinear history.

The virtual work of an element of length l now follows, as

$$\int \sigma_{zz} \delta\varepsilon_{zz}\, dV = \int_l \int_A [w'_0 - yv'' + \tfrac{1}{2}(v')^2] E [-y\delta v'' + v'\delta v']\, dA\, dz.$$

NONLINEAR ANALYSIS

Noting that AEw'_0 is the axial tensile force P, and that $\int y\, dA = 0$, this becomes

$$\int_l \left[(Pv'\delta v' + EIv''\delta v'' + \tfrac{1}{2}A(v')^3 \delta v' \right] dz.$$

But we are admitting only 'moderately large' rotations so the last term is negligible. Putting $v(z) = \mathbf{N}\mathbf{d}_g$ as usual the virtual work becomes

$$\bar{\mathbf{d}}_g^t \mathbf{k}_g \mathbf{d}_g = \bar{\mathbf{d}}_g^t \left[P \int_l \mathbf{N}''^t \mathbf{N}' \, dz + EI \int_l \mathbf{N}'''^t \mathbf{N}'' \, dz \right] \mathbf{d}_g.$$

The last term is the usual elastic stiffness term $\mathbf{k}_E$ (4.8), whilst the first term is the geometric stiffness $P\mathbf{k}_G$, which using (4.7) again is readily evaluated as

$$\mathbf{k}_G = \int_l \mathbf{N}''^t \mathbf{N}' \, dz = \frac{1}{30l} \begin{bmatrix} 36 & 3l & -36 & 3l \\ & 4l^2 & -3l & -l^2 \\ & & 36 & -3l \\ \text{symm} & & & 4l^2 \end{bmatrix}. \tag{11.9}$$

Plate bending elements, solid bricks and so on can all be treated in a similar fashion. The geometric stiffness will always take a similar form involving an integral of the current stresses over a product of the shape functions. The net element tangent stiffness will always appear like the above in the form (c.f. equation (11.8))

$$\mathbf{k}_T = \mathbf{k}_E + P\mathbf{k}_G. \tag{11.10}$$

Remember that this is for moderate rotations, and $\mathbf{k}_T$ refers to local co-ordinates (y, z) which may have rotated significantly with respect to the fixed global axes. The element stiffness referred to fixed axes is however readily found using the usual transformation $\mathbf{T}^t\mathbf{k}\mathbf{T}$ (4.13).

The separate contribution to $\mathbf{k}_T$ from $\mathbf{k}_E$ and $\mathbf{k}_G$ will add or subtract depending on the sign of the stress field 'P'. If this is negative (compressive) the structure clearly has lost some net stiffness. It is possible then to discuss the *initial buckling* of structures without resorting to a full incremental solution. The assumption is made that if a slender structure like a strut or flat plate has its compressive load increased there will come a time when the displacements suddenly increase dramatically even when

A FINITE ELEMENT PRIMER

the other loads **R** are trivial. That is the solution to $\mathbf{k}_T \delta \mathbf{r} = \delta \mathbf{R}$ becomes singular, or the determinant of $\mathbf{K}_T$ is zero,

$$|\mathbf{k}_E - P\mathbf{k}_G| = 0. \quad (11.11)$$

This equation gives the value of P at which initial buckling will occur, in the absence of initial imperfections. If standard overstiff displacement-based elements are used, the predicted buckling loads will be too high. The reader may care to verify the buckling load for a simply supported strut by inserting $\mathbf{k}_G$ (11.9) and $\mathbf{k}_E$ (4.8) into (11.11) after deleting the 1st and 3rd rows and columns ($r_1 = r_3 = 0$). (The result is $P = 12EI/l^2$ which is 21% too high.)

For realistic idealisations involving many degrees of freedoms **r**, equation (11.11) is a standard eigenvalue problem which all finite element systems can handle, and the smallest eigenvalue is the initial buckling load. Most systems use an iterative routine for solving eigenvalues and the smallest one is delivered first. It should be explained that the initial buckling load is derived by assuming that the 'large' deflections are still moderate. After initial buckling some structures like struts effectively lose their stiffness completely, some like stiffened plates may be further loaded at a reduced stiffness, and some like shallow arches or shells may snap through and recover their stiffness completely. A full incremental solution is necessary to investigate their behaviour.

11.3 Incremental Solutions

The equation $\mathbf{K}_T \delta \mathbf{r} = \delta \mathbf{R}$ must be used to build up a nonlinear history and some form of incremental analysis is therefore needed in which the loading vector **R** is applied as a series of small increments $\delta \mathbf{R}$, and the corresponding displacements $\delta \mathbf{r}$ are solved at each step. There are many ways of implementing this incremental process, and only a few of the most used are described briefly. None is perfect, and the appropriate choice depends on the type of structure and the current behaviour of the load ~ deflection curve.

The important thing to note is that the geometric stiffness depends on the current stresses and the current position of the element. Likewise the elastic stiffness $\mathbf{K}_E$, although it remains constant for small strains, will alter if the displacements ($\mathbf{r} = \sum d\mathbf{r}$) cause significant rigid body movement of the element ($\mathbf{K}_E$ would become $\mathbf{T}^t\mathbf{K}_E\mathbf{T}$ (4.13)). Thus it is the *tangent* stiffness matrix $\mathbf{K}_T$ which will be constructed rather than the current *secant* stiffness, which relates total displacements to total forces.

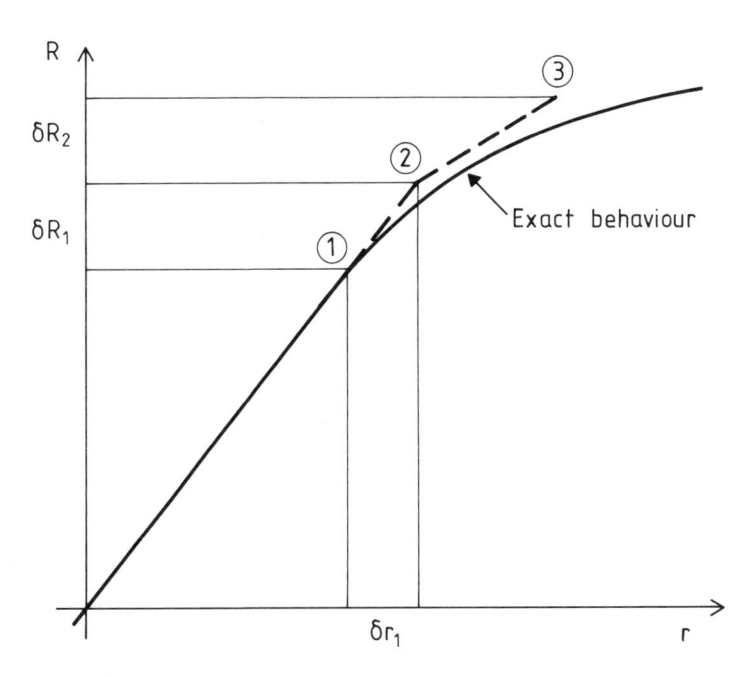

Figure 11.4 Incremental drift.

The straightforward approach is illustrated in Figure 11.4. Point (1) is a starting point on the $\mathbf{R} \sim \mathbf{r}$ curve where it is assumed we known the current displacement $\mathbf{r}$, and local stress field σ. Since $\mathbf{K}_T$ is a function of both $\mathbf{r}$ and σ we can apply a small increment in load $\delta \mathbf{R}$, use $\mathbf{K}_T(1)$ to obtain $\delta \mathbf{r} = \mathbf{K}_T^{-1} \delta \mathbf{R}$, and thence a new $\mathbf{K}_T(2)$ using the displacements and stresses updated by $\delta \mathbf{r}_1$, and $\delta \sigma$. However although $\mathbf{K}_T(2)$ may be correct, the point (2) may not be on the $\mathbf{R} - \mathbf{r}$ curve so the launch to the point (3) will be somewhat above a curve like Figure 11.4, and the drift away may continue. It is therefore usual to apply an equilibrium correction at point (2). We can evaluate the nodal forces at (2) from the current element stresses, using the PVD, as:

$$\delta \mathbf{P}_g^t \bar{\mathbf{d}}_g = \int_V \delta \sigma^t \bar{\varepsilon} \, dV = \int_V \delta \sigma^t \mathbf{B} \, dV \bar{\mathbf{d}}_g$$

or

$$\delta \mathbf{P}_g = \int_V \mathbf{B}^t \delta \sigma \, dV. \tag{11.12}$$

A FINITE ELEMENT PRIMER

These small forces can be subtracted from $\delta \mathbf{R}$ and the process repeated with the new $\mathbf{K}_T$ as shown in Figure 11.5a.

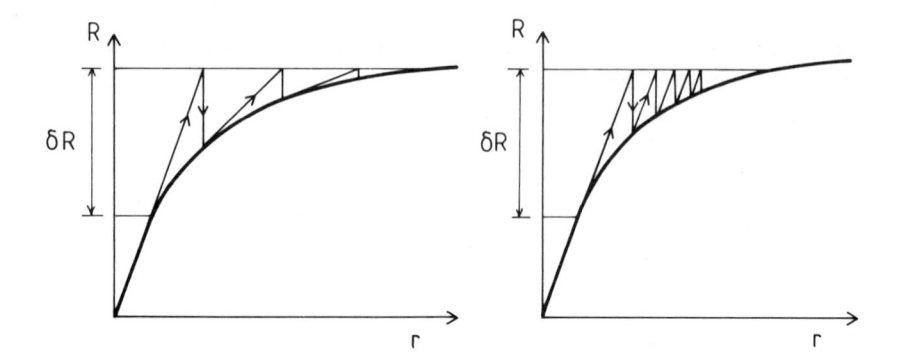

Figure 11.5 Newton–Raphson.

Figure 11.5b Modified Newton–Raphson.

It is expensive to evaluate $\delta \mathbf{r} = \mathbf{k}_T^{-1} \delta \mathbf{R}$ from several updates on $\mathbf{k}_T$ so it is usual to keep $\mathbf{k}_T$ constant and accept that the number of iterations will be more – as shown in Figure 11.5b. This process can take some time if the structure is near to collapse through buckling or plastic flow when $\mathbf{K}_T$ will approach the singular value of a mechanism. It can fail completely when applied to a hardening curve as shown at the point (1) in Figure 11.6, whereas the unmodified Newton–Raphson strategy (2) will still work.

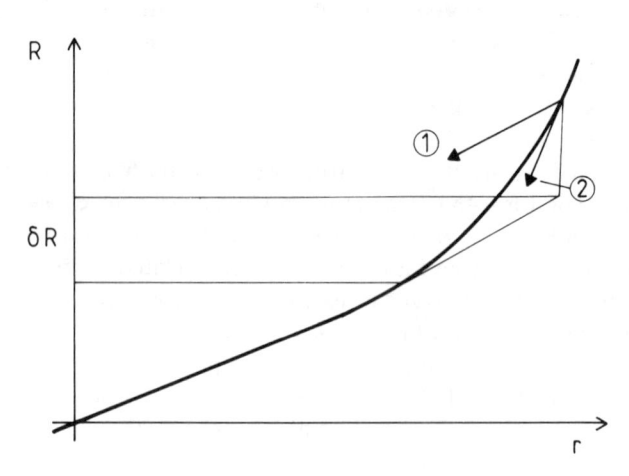

Figure 11.6 Hardening Structure.

Another trick is to apply a further load **R** together with only the first equilibrium correction to cancel the unloading. Yet another ruse is to go back to the beginning of the step and use the mean $\mathbf{K}_T = \frac{1}{2}(\mathbf{K}_T(1) + \mathbf{K}_T(2))$ to forecast $\delta \mathbf{r} = \mathbf{K}_T^{-1} \delta \mathbf{R}$. This will work whether the curve is softening or hardening, and is exact if the current $\mathbf{R} \sim \mathbf{r}$ relationship is quadratic. Whatever iterative numerical technique is used, the residual forces of (11.12) are a valuable check on the convergence of the process, and should be available as such to the user of a finite element system.

The need to form a changing stiffness matrix and then solve a set of equations each load increment can be avoided completely by presenting the problem as a dynamic one. The full load **R** is applied immediately and the subsequent nonlinear motion is followed, with a little damping introduced to ensure that the static solution is achieved after a decent time. The dynamic progress is constructed using the explicit time-marching method discussed after (10.15) in which the stiffness matrix is not inverted. This 'dynamic relaxation' method can be competitive and has the virtue that it will solve snap-through problems like the shallow arch where $\mathbf{K}_T$ may be temporarily singular.

Proprietary finite-element systems are not always straightforward to use, and many decisions have to be taken by the user on initial conditions, step sizes, convergence, numerical damping, over/under relaxation, equilibrium checks, monitoring results and so on. These methods can be abused, and are always expensive. The descriptions in this Primer are very basic and any novice would be well advised to seek advice from an experienced user, the finite-element system developer, or NAFEMS.

11.4 Inelastic Material Behaviour

The other major form of nonlinearity occurs in the material's constitutive stress-strain law once the stresses go beyond the linear elastic range. The finite element strategy then is identical in most cases to the previous section in that $\mathbf{K}_T$ will be found directly and updated using current forms of the stress-strain law. We concentrate on this aspect but mention that there are alternatives in which a local element is simply deemed to have failed at a certain strain and lost some component of its tangent stiffness thereby depleted the global stiffness. This approach is possible for a highly ductile material like mild steel, which can be considered as yielding completely at a known yield stress, provided the stress-field is unidirectional as in slender bars and beams. Yet again some materials

A FINITE ELEMENT PRIMER

may fracture and create gaps between elements. Physical gaps may appear and disappear in contacting surfaces, and special elements and tactics are available in some systems for this form of nonlinear behaviour.

Turning now to the general case where nonlinear behaviour progresses continuously after a linear elastic phase, we consider first the unidirectional case.

Most materials will have a linear range, $\varepsilon = \sigma/E$, for moderate strains and thereafter the material tangent stiffness E_T will decrease in some fashion as shown in Figure 11.7. This behaviour will be quantified by experiment, by code of practice, by the manufacturer's data sheets, or from the open literature. For simplicity it is not uncommon to assume that the tangent modulus has a constant reduced value after some initial yield stress.

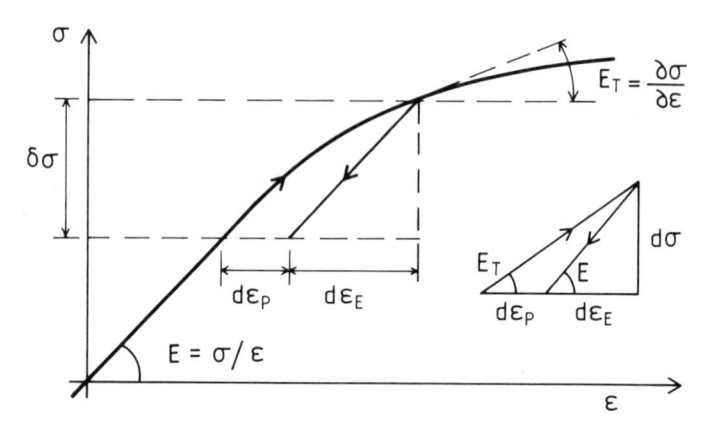

Figure 11.7 Inelastic material.

The incremental (Prandtl–Reuss) theory of plasticity identifies two strain components, the elastic recoverable part ε_E and the plastic permanent part ε_P. There are therefore three material moduli involved in the stress increment shown in Figure 11.7

$$E = \frac{d\sigma}{\partial\varepsilon_E}; \quad E_P = \frac{\partial\sigma}{\partial\varepsilon_P}; \quad E_T = \frac{d\sigma}{d\varepsilon_E + d\varepsilon_P}.$$

The plastic flexibility is consequently given by

$$\frac{1}{E_P} = \frac{1}{E} - \frac{1}{E_T}. \tag{11.13}$$

For a unidirectional stress system it is only necessary to insert the tangent modulus E_T into the element stiffness (5.4) to produce an element tangent stiffness $\mathbf{k}_T$. A uniformly strained bar will simply have its elastic stiffness factored by E_T/E. A beam element has a linearly varying strain field and its moment curvature relationship may have to be constructed or simplified.

In a complex two dimensional or three dimensional stress field we have to broaden the concept of yield and tangent moduli. The 'yield' criteria must depend on a reference or 'equivalent' stress $\bar{\sigma}$ which is invariant to choice of reference system (material failure is no respecter of x, y or z). The Prandtl–Reuss flow rule stipulates that the permanent plastic strain components produce no change in volume. We therefore pose these requirements in terms of *stress deviators* $\mathbf{s}$ which depend on the shear stresses and on the excess of the direct stresses over a mean (hydrostatic) stress

$$\sigma_m = 1/3(\sigma_{xx} + \sigma_{yy} + \sigma_{zz}).$$

This implies that no plastic flow occurs due to σ_m alone. The most convenient vector to use is

$$\mathbf{s}^t = [\sigma_{xx} - \sigma_m, \sigma_{yy} - \sigma_m, \sigma_{zz} - \sigma_m, \sqrt{2}\sigma_{xy}, \sqrt{2}\sigma_{yz}, \sqrt{2}\sigma_{zx}]. \quad (11.14)$$

We find that the (Von Mises) reference stress is

$$\bar{\sigma}^2 = \tfrac{3}{2}\mathbf{s}^t\mathbf{s}. \quad (11.15)$$

Other reference stresses are used, but this one seems to represent fairly well the behaviour of ductile steels and aluminium alloys.

The incremental plastic strain vector is given by

$$d\varepsilon_P = \frac{3\,d\bar{\sigma}}{2E_P\bar{\sigma}}\mathbf{s}.$$

Substituting into $d\varepsilon = d\varepsilon_E + d\varepsilon_P$ and inverting we find that

$$d\boldsymbol{\sigma} = (\mathbf{E} - \mathbf{E}_P)\,d\mathbf{E} \quad (11.16)$$

where

$$\mathbf{E}_P = \frac{9G^2 \mathbf{s}\mathbf{s}^t}{(3G + E_P)\bar{\sigma}^2}. \quad (11.17)$$

A FINITE ELEMENT PRIMER

This can be thought of as the material plastic stiffness. Before yielding takes place ($\bar{\sigma} < Y$) when $E = E_T$, $E_P = 0$. It is also zero if an increment of applied load results in *unloading* ($d\bar{\sigma} < 0$) which is assumed to occur elastically.

On substituting (11.16) into the element work integral we find the tangent stiffness as

$$\mathbf{K}_T = \mathbf{K}_E - \mathbf{K}_P \tag{11.18}$$

where

$$\mathbf{K}_P = \int_V \mathbf{B}^t \mathbf{E}_P \mathbf{B} \, dV. \tag{11.19}$$

Thus the modification to the elastic stiffness is similar to that brought about by gross deformations, indeed both may occur simultaneously

$$\delta \mathbf{R} = (\mathbf{K}_E + \mathbf{K}_G - \mathbf{K}_P) \delta \mathbf{r}.$$

Programmes need the data to construct $\mathbf{E}_P$, either in the form of stress, tangent modulus, or plastic modulus as a function of strain.

In many cases the user will wish to find the maximum load at collapse ($\mathbf{K}_T$ singular) but this is computationally futile and often we stop the loading at some unacceptable deformation rather than chase the large deflections of an increasingly floppy structure. Plasticity can be a very local effect and may unload in one region to appear somewhere else. If the structure is loaded cyclically with plastic deformation at each stage then 'shakedown' may occur and there are special techniques for predicting the limit of this process.

The element stiffness integrals themselves can be expensive to update as plasticity progresses with increasing load. Higher-order elements will have plastic zones within them and it is invariably the practice to rely on the Gauss Point stresses to represent the local nature of the element. This is not strictly accurate when the stress-strain field changes its nature (it may actually be discontinuous) and the Gauss points cease to be the correct sampling points. For this reason, many practitioners, and a few finite element systems, will restrict inelastic analysis to constant strain triangles or tetrahedra, with single measures of stress and volume.

The complete nonlinear field, whether due to gross deformations, plasticity, or both, is a complex problem. Some of the codes are not user-friendly and demand much of the inexperienced user. They are also expensive to use, but usually the cost is in the number of iterations rather than the number of unknowns. There is likely therefore to be a trend to writing nonlinear software specifically for microcomputers or super micros.

This chapter has been but a brief introduction to nonlinear finite elements. At the best we hope we have explained the principles and at least, given the novice a better chance of understanding the manuals!

12. Modelling

12.1 Introduction

The successful use of finite element programs relies upon the skill of the analyst in setting up the model for the problem. This requires that he has a knowledge of, or an instinct for, the load paths through the structure and the resulting stress distributions so that he can define a suitable mesh. In addition he must also have more than a passing appreciation of the theory that was used to formulate the types of elements that are being used. Such a knowledge, and how the structural behaviour and the element behaviour interact, requires a considerable degree of experience on the part of the user which can only be obtained by practical use of the method. This must also be backed up by the analyst actively reviewing the solution which is obtained and comparing it with the expected results. This requires that time is spent in inspecting and understanding the output produced by the program and allowance must be made for this when estimates are being made for the time schedules of the job. It is at this stage of gaining an overall appreciation of the structural behaviour when carrying out order of magnitude checks that most silly errors (and even the most experienced user make them) are discovered.

12.2 Basic Modelling Considerations

The basic perceived nature of the finite element method can be very misleading regarding the types of problems that can be solved using the method. At first sight it appears that a series of node points are defined and that therefore point information is all important, with the solution being accurate at the nodes. In fact it was shown in chapter 3 that this is not the case. The essence of the finite element method, especially for those elements that are defined by assumed shape functions, is to produce a solution that is correct on average across an element but which might have a considerable error for any point on the element, including node points. The smearing of the behaviour over the element has great consequences regarding the problems that are most suitable for analysis by the finite element method and those that are not. It is most accurate for continuum problems where the geometry, the material properties and the loadings all change in a continuous manner over the field that is being modelled. If there are discontinuities in any quantity then they are smeared and the modelling of such discontinuities is not precise. Obviously the mesh can, in theory, always be made smaller to better acccommodate any sharp changes in the parameters but there will always be a physical limit to the degree of refinement which is possible. This is especially true for three dimensional problems where a change in the mesh density in one region of the structure will have a consequential effect on the mesh everywhere within the model, creating problems in generating the mesh. For three dimensional problems it is usually impossible to refine the mesh uniformly in all directions throughout the structure since the cost of the analysis rises at a rate that is likely to be proportional to the cube of the number of elements. This is such a large rate of change that even a relatively coarse three dimensional mesh is expensive to run.

12.3 Modelling Considerations for Linear and Non-linear Problems

There are two common forms of non-linearity that arise in structural analysis, large deflections and plasticity. In a linear analysis the deflections can be large or stresses can exceed yield without affecting the cost of the analysis since it ignores their effect upon the solution. If the solution is such that the regions where the non-linearities are predicted to occur are localised and they do not coincide with the regions of interest for the analysis then the user is normally at liberty to ignore the non-linear behaviour. If the results are such that non-linearities will occur throughout the structure or within regions that are of direct interest to the analyst then he should either consider the structure to have failed or be prepared to carry out a non-linear analysis. If a non-linear analysis is being conducted then the user must pay considerably more attention to the details of the model, even for those regions that are not of direct interest. The non-linear analysis will treat all occurrences of the non-linear behaviour that is being investigated with equal importance. If some regions of the model trigger the non-linear analysis due to poor modelling rather than it being a real structural effect then a considerable computer cost will be incurred with few useful results being generated. Such modelling problems can arise due to poor representation of the loadings or a poor mesh definition where the load is applied. Typically, a point load applied directly to a solid element will cause the model to attempt to reproduce an infinite stress under the applied load and this will trigger non-linear behaviour. A similar effect will occur for poor support modelling. The user must also pay close attention to the definition of the mesh for non-linear problems. Typically, in a well designed joint there will be no load offsets but for plate or beam models, if the connections are not modelled correctly, large deflections can be predicted where they should not arise. Similarly, if the actual joint does have offsets but these are not included in the model then the non-linear behaviour will be underestimated.

If the user is intending to carry out a non-linear analysis then it is imperative that a linear analysis, or a series of linear analyses are conducted first. The results of a non-linear analysis can be very sensitive to the actual model used and if this is not truly representative of the real structure then the solution will have no physical meaning. The types of non-linear response, the regions of the structure that are non-linear, and non-representative idealisations, can all be discovered from the linear analysis. This is all important since any non-linear analysis will require the solution to be found by a series of iterations and these will invariably be expensive. Such considerations become even more important if the user is contemplating a non-linear dynamic analysis which require iterations on top of time stepping. Before the user starts such a job he should be very sure about the purpose of the analysis and be prepared to pay very big computing bills.

12.4 Modelling Joints

One consequence of these considerations is that it is much easier to devise a model for a complicated but smooth three dimensional geometry than it is to model accurately a joint within a framework. In practice the effect of joints is probably the most difficult thing to model within a finite element analysis. The method usually connects nodes across elements. If the effects of a joint are to be introduced then the nodes are not connected directly and some form of joint stiffness must be defined. This immediately increases the number of degrees of freedom in the model and usually the stiffness of the joint is such that it introduces a large disparity in the stiffness across the interface, with the possibility of consequential rounding errors in the solution process. Even worse than this, from the point of view of the accuracy of the modelling, is the fact that the element produces interface forces that are kinematically equivalent and the distribution of these can often bear little resemblance to the interface pressures that the joint will experience. This arises because of the inherent smearing of the finite element method and the lack of any guarantee of point wise convergence. As a consequence, any joint model that is not kinematically consistent with the associated elements is likely to be only very approximate. Such a problem occurs with bolted joints, where the bolts are not closely spaced, or with joints that are spot welded. In this case there are point connections between elements and to model such effects correctly a fine mesh is required in the region of the discontinuity introduced by the joint. This degree of refinement is rarely carried out and the user must appreciate that he is not modelling a point connection as he intended but is, in fact, smearing the effect of the joint in some manner defined by the element shape functions along the element edge. If the analyst is attempting to model the opening and closing of a joint then great care must be taken to ensure that the model is realistic. Even more uncertainty is introduced if there are bearings in the system. It is a reasonable assumption that there is no stiffness in the direction of free movement, but the degree of stiffness transverse to this direction can be very debatable. The analyst is usually left with the choice, for the want of any better information, of assuming either that there is a rigid fixity or that it is free. The problem of the modelling of joints can be most important for a dynamic analysis. They are usually designed to carry the working loads so that, for a static analysis, the main load path is through the strongest direction of the joint. However, for dynamics, the solution process is such that the lowest modes can follow load paths corresponding to the lowest stiffness, that is they tend to activate joints in their weakest, most ill-defined, directions. This can lead to serious discrepancies between the computed and the measured dynamic response.

Glued or other continuous joints, or those with closely pitched rivets or bolts can be modelled in one of two ways. The actual jointing can be assumed to be very stiff, in which case, the joints are essentially ignored and corresponding elements are just assembled together in the usual manner. Alternatively, a special finite element representing the joint can be defined and used in the analysis. This is valid for those cases where the pitch of the jointing material is so close that its effects can be smeared into a continuum which is then amenable to a finite element idealisation. Often this can be done by defining appropriate values for the elastic properties of standard elements.

12.5 Modelling Offsets

A modelling problem that commonly occurs with beam, plate or shell

A FINITE ELEMENT PRIMER

elements arises because these elements are defined in terms of the geometry of the neutral surface of the element. The thickness of the plate or beam is only included indirectly, the assumption being that the element is very thin. In practice the thickness of the structure can be significant. This is illustrated in Figure 12.1, where two beam elements connect at right angles.

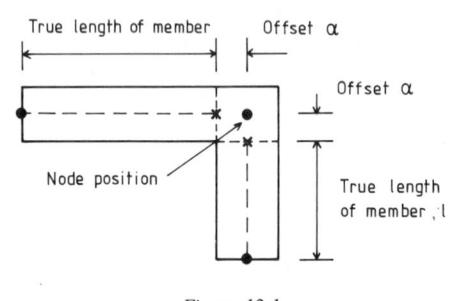

Figure 12.1

There are various ways in which this can be modelled. If the beam offset α is less than 1% of the beam length l then the offset of the two centre lines can be ignored as being insignificant. If α is greater than $0.25\,l$ then the user should consider seriously whether a beam model is relevant, since the aspect ratio of the beams are such that the assumptions of beam theory are becoming debatable. For the intermediate cases the user is left with having to model the offsets in some way. One solution that is always possible is shown in Figure 12.2, where the beams are modelled up to the junction and are then connected by a fictitious rigid element.

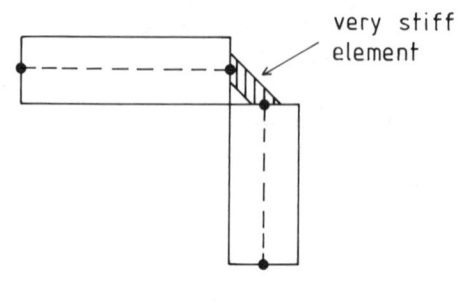

Figure 12.2

MODELLING

This is given large values of cross-sectional area and moments of inertia so that it models the rigidity of the joint. Although such a solution is always possible it is not very satisfactory since the user has to make an arbitrary choice of section properties for the fictitious element and, more importantly, the high stiffness that this induces can give rise to conditioning problems within the solution. There will always be inherent conditioning problems for structures that involve stretching and bending and the inclusion of stiff elements makes this worse. This type of modelling is definitely not recommended and should only be used as a last resort. The alternatives that do not lead to such difficulties involve the introduction of rigid connections, either in the form of rigid offsets or as multi-point constraints. The user is then restricted by what is available within the program that he is using. Probably the best solution is to use elements that allow rigid offsets to be defined so that the node need not be at the end of the element. This is illustrated in Figure 12.3.

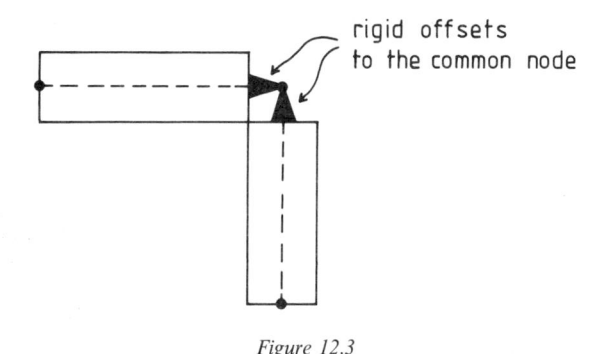

Figure 12.3

The effect of the rigid offset is obtained by including an appropriate transformation matrix in the formulation of the element. It does not require the user to define any arbitrary high stiffness. The junction can then be modelled as shown in Figure 12.3, where the rigid offsets are used to place the structural node at the junction of the two bars but allows their correct length to be used in the model.

The same result can be obtained by means of constraint (or multi-point constraint) equations. In this case the model is defined as shown in Figure 12.4, where the ends of the two beams are not connected. Constraint equations are then defined to connect these freedoms together to simulate a rigid connection. Again the user does not have to define any arbitrary

A FINITE ELEMENT PRIMER

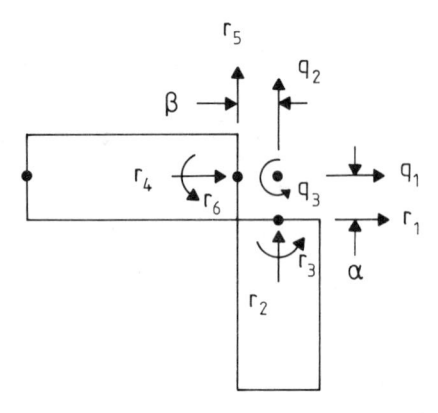

Figure 12.4

high stiffness but he does have to define the constraint equations that connect the two nodes. These can be written in one of two forms:

$$\begin{bmatrix} r_1 \\ r_2 \\ r_3 \\ r_4 \\ r_5 \\ r_6 \end{bmatrix} = \begin{bmatrix} 1 & 0 & -\alpha \\ 0 & 1 & 0 \\ 0 & 0 & 1 \\ 1 & 0 & 0 \\ 0 & 1 & -\beta \\ 0 & 0 & 1 \end{bmatrix} \begin{bmatrix} q_1 \\ q_2 \\ q_3 \end{bmatrix} \quad \text{or} \quad \begin{matrix} -r_1 + \alpha r_3 + r_4 = 0 \\ -r_2 + \beta r_3 + r_5 = 0 \\ -r_3 + r_6 = 0 \end{matrix}$$

The modelling of axisymmetric thin shells gives rise to exactly the same problems except that the correct modelling of the junction is even more important in this case since it is the satisfaction of continuity at such junctions which gives rise to the local discontinuity stresses. The determination of the stresses that arise at such discontinuities is often one of the main objectives of the analysis. Since such stresses die away with distance from the junction their magnitude can be seriously affected by the effective length of the shell if its other end is within this die away length. If rigid offsets or constraint equations are used for the shell junction analysis then the stiffness of the material at the junction should be included by means of a point element, especially if there is a flange at this point. Rigid offsets of this type are also used to connect plate and shell elements to the centre line of offset reinforcing beams. This is illustrated in Figure 12.5.

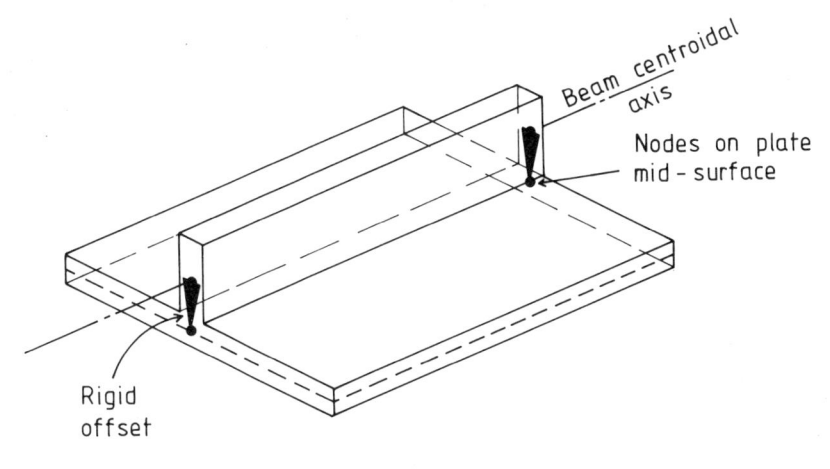

Figure 12.5

When using constraint equations with axisymmetric shells it is important that the user remembers that the shell is in fact axisymmetric and not just the cross sectional slice that is usually presented as the structure plots. If care is not taken the hoop growth can be constrained making the model very overstiff.

12.6 Modelling of Supports

The majority of real structures are supported in some way and these supports must be included in the analysis. There are a few occasions where the structure is not supported, for example an aircraft in flight. In this case the loads must be self equilibrating so that rigid body motions do not occur. The structure can then be supported with any arbitrary set of statically determinate set of supports. The displacements that are found will be relative to the planes defined by the choice of these supports and the calculated deflections will vary with the choice of supports. However, the different solutions obtained by choosing different sets of supports only vary by rigid body movements which means that the stresses that are calculated will be the same for any set of statically determinate supports.

The situation is very different for dynamic problems. The unsupported free-free vibration case occurs quite frequently and, in this case, the analysis must be conducted without any supports. If there are m rigid body motions then there will be m zero frequency modes with associated

A FINITE ELEMENT PRIMER

mode shapes that are the rigid body movements. All other modes will cause deformation of the structure and will have non-zero frequencies. The eigenvalue extraction schemes used by some programs cannot cope with singular stiffness matrices and if the analyst requires to find the eigenvalues and vectors of a structure with rigid body motions then he should check, possibly by running a simple test example, that the extraction method that is to be used is valid. If it is not valid the structure can be supported on a set of very flexible supports, in which case the resulting frequencies will not be zero but they will be small compared to the non-zero modes. The vectors will still be the rigid body motions.

The majority of structures do have a relatively well defined set of supports, although, even in simple cases, some thought must be given to the definition of the supports within the model. Typically a portion of the structure might be the cantilever shown in Figure 12.6a, where this is modelled using membrane elements.

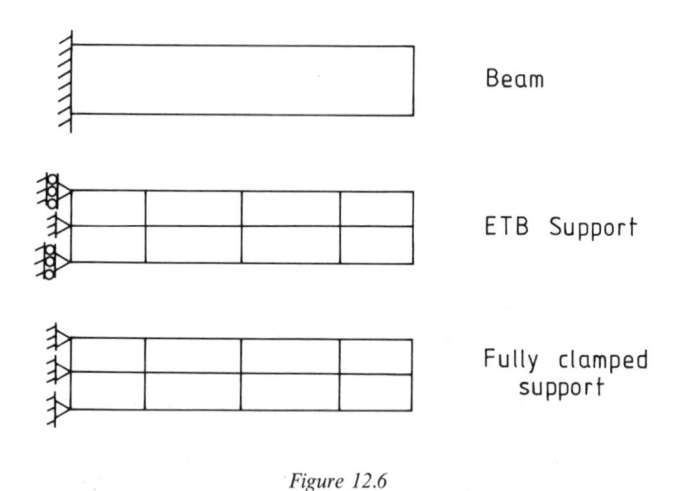

Figure 12.6

The two possible representations of the built in end are shown in Figures 12.6b and 12.6c. In Figure 12.6b the supports are such that plane sections remain plane but no transverse stresses are induced. This corresponds to the assumptions within Engineers Theory of Bending. Figure 12.6c has the cantilevered end fully built in. In this case the transverse strains at the fixed end of the beam suppressed which, in turn, induces transverse stresses. These can have a significant stiffening effect, depending upon the

MODELLING

aspect ratio of the beam, and the user must choose which case is more relevant to his analysis. As the aspect ratio of the beam gets higher this effect is less important.

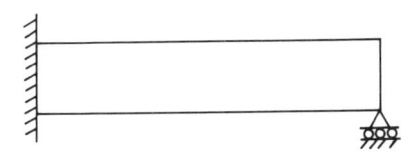

Figure 12.7

The user must take care where there is a support prop, as shown in Figure 12.7. If this is modelled as shown then it will give rise to a stress concentration at the support. For the real structure there will be a reinforcing frame or some form of loading pad at the support point. If the stresses around the support point are of interest then the details of the support must be included in the finite element model. If the stresses at the support are not of interest then the stress concentration that arises from the unrealistic modelling can be ignored. This is only true for a linear analysis. For a non-linear problem a poor idealisation is likely to lead to an apparent failure at the support and the results obtained will be of little use.

The consequences of the chosen set of supports can be rather more subtle than this. Consider the axisymmetric disc shown in Figure 12.8a.

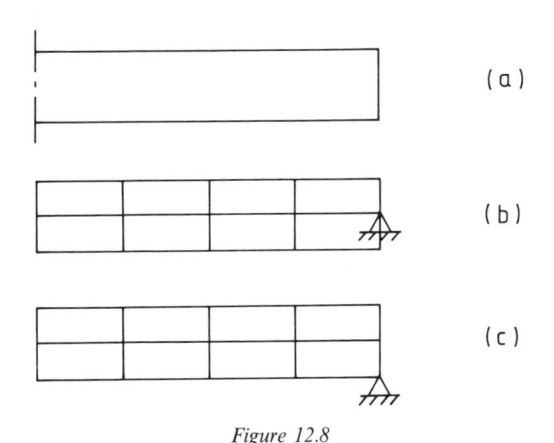

Figure 12.8

241

A FINITE ELEMENT PRIMER

The pinned end support will give two different results for the fixities shown in either Figure 12.8b or 12.8c. In the first case the pin is assumed to be on the centre line of the plate and no stretching of the mid-surface will occur. In Figure 12.8c the pin is at the lower corner and this will induce mid-plane stretching and will lead to a considerably stiffer behaviour of the plate.

There are considerable modelling problems for those structures that rest upon an elastic foundation. This is typified by a concrete dam built onto a relatively hard bedrock. Here the material properties of the concrete and the foundation can be comparable and it is very difficult to decide where the boundary of the model should be taken. So far as the structure is concerned the stresses will be greatest if the foundation is assumed to be rigid. If the displacements are important this will have the effect of underestimating them. If the structure and the foundation have comparable stiffness then they should both be included in the model. There are two conventional methods for doing this. Firstly the structure can be supported on a series of springs. Obviously, a problem that the user has to resolve here is the choice of the spring stiffness to use. This is usually found either from an analytical half space model of the foundation alone or from some empirical equation. Alternatively, a part of the foundation can be included in the finite element model of the structure. The amount of foundation that needs to be included can be found by carrying a preliminary analysis of the foundation alone, applying unit loads at points corresponding to the structure connections and finding out the distance where the effects of these have decayed to a small value. This is used to define the required extent of the foundation component of the model. The effect of layering in the foundations can be found by a similar preliminary analysis.

12.7 The Use of Constraint Equations

Wherever possible the analyst should simplify the idealisations that are used within the model so that areas of uncertainty (typically joints or very stiff or flexible sections of structure) are modelled using one of two extreme assumptions, either that it has no stiffness or that it is completely rigid. The analyst should avoid the obvious temptation of trying to model the actual stiffness of such regions for various reasons. Firstly the value of the stiffness is not usually known and the consequences of guessing values will be unpredictable. Secondly, the detailed mesh that is required to model the area is usually uneconomic compared to the overall cost of the analysis. Finally, and most importantly, any such areas will almost invariably have stiffness values that are orders of magnitude different from the rest of the structure and this will lead to ill-conditioning of the equations. Within this area the analyst should avoid using very short or very long elements, a very high or low Young's modulus, non-realistic geometrical parameters such as thickness or moments of inertia, or elements of a different type from that used in the rest of the model. Where it is applicable the simplest extreme assumption is to assume that the area involved has no stiffness at all and it is omitted from the model. If a joint is involved then nodes at either end of the joint are left unconnected. In this case the user must check that either the structure is supported against rigid body motions and contains no mechanisms or that the applied loads are self equilibrating. The other modelling extreme that is often used is to assume that the area of interest is rigid, in which case nodes involved are then fully connected. The simplest form of this is the rigid offset problem discussed in section 12.5. There are more general forms that can arise, typically that plane sections remain plane, that a part of the structure is rigid but it can move as a rigid body, or that incompatible elements or meshes have to be joined in some rational manner.

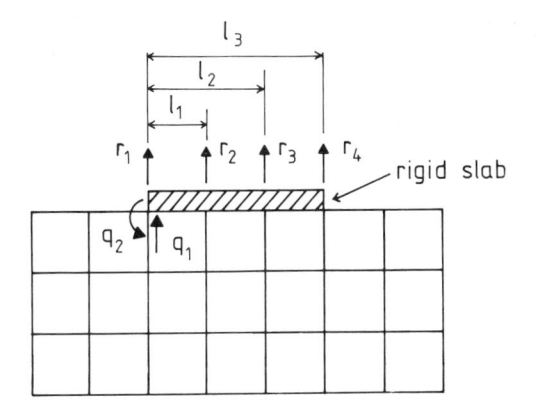

Figure 12.9

Constraint equations define a relationship between sets of displacements within the mesh. They can be expressed in various ways and different programs reflect these variations by the manner in which the constraint equations are implemented. To illustrate the possibilities consider the geometry of Figure 12.9 where a rectangular region has a rigid slab on top of it. For simplicity it is assumed that this slab constrains the vertical displacements r_1 to r_4 so that they remain straight but it has no effect upon any other freedom. The constraint can then be expressed in terms of the two possible displacements of this line of nodes, a vertical translation, q_1 and a rigid body rotation, q_2. These are shown in Figure 12.9. The rotational freedom q_2 can be taken about any point, not necessarily a node point. For this example it has been taken as the node where r_1 occurs. The freedoms q_1 and q_2 can then be used to define the constraint transformation

$$\mathbf{r}_c = \begin{bmatrix} r_1 \\ r_2 \\ r_3 \\ r_4 \end{bmatrix} = \begin{bmatrix} 1 & 0 \\ 1 & l_1 \\ 1 & l_2 \\ 1 & l_3 \end{bmatrix} \begin{bmatrix} q_1 \\ q_2 \end{bmatrix} = \mathbf{cq}$$

where $\mathbf{r}_c$ is the portion of the displacement vector, $\mathbf{r}$, that contains r_1 to r_4. Calling the remaining displacements $\mathbf{r}_f$ then the full set of displacements can be written as

$$\mathbf{r} = \begin{bmatrix} \mathbf{r}_f \\ \mathbf{r}_c \end{bmatrix} = \begin{bmatrix} \mathbf{I} & 0 \\ 0 & \mathbf{c} \end{bmatrix} \begin{bmatrix} \mathbf{r}_f \\ \mathbf{q} \end{bmatrix} = \mathbf{Tr}'. \tag{12.1}$$

The coefficient matrices (stiffness matrix or mass matrix) can then be modified to include the constraints by forming the usual product

$$\mathbf{k}' = \mathbf{T}^t \mathbf{k} \mathbf{T}.$$

The equations that are to be solved are then

$$\mathbf{k}'\mathbf{r}' = \mathbf{T}^t \mathbf{R} = \mathbf{R}'.$$

Once this has been solved then the displacements r_1 to r_4 can be recovered from the constraint definition of equations (12.1).

This implementation requires that the constraint equations are defined before the stiffness matrix is factorised since the matrix itself is modified (conceptually it can be considered that the constraints are assembled into the equations). The number of equations that have to be solved is then reduced by a number related to the number of constraints. This method of applying the constraints will tend to improve the conditioning of the equations compared to the unconstrained set. There are also deficiencies associated with this form of constraint application. It will generally tend to increase the bandwidth of the equations, sometimes a considerable increase can occur. Since the constraints are applied before assembly it also means that if the constraints are changed within a run then the constrained stiffness matrix must be reformed and refactored and this can be an expensive process. If a dynamic problem is to be solved then the same constraints must also be applied to the mass and, where it has been formed, the damping matrix. It is also necessary to transform the forces so that the correct load vector is formed and when the equations have been solved then the displacements must be recovered from the constrained system. These transformations are usually carried out internally within the program and the user need not consider them.

An alternative method of applying the constraints is to consider the forces required to satisfy the constraints to be Lagrange multipliers. These can then be found as a part of the solution process. In practice the method is implemented by solving the equations without any constraints and then solving a subsidiary set of flexibility equations that modify the solution so that the constraints are satisfied. The advantages of this form are that the original set of equations are retained so that the bandwidth is not changed by the constraints and, secondly, the constraints can be changed at any time or with any load case without having to re-factor the stiffness matrix. Also the load vector does not have to be modified for the constraints and the solution gives the displacements directly in the correct coordinate system. The disadvantages are that the subsidiary flexibility matrix that must be solved to enforce the constraints will be fully populated and, if there are many constraints, then it will be large, leading to an expensive solution. Since the constraints are being applied after the stiffness matrix has been factorised there can be problems with this form of constraint implementation that arises from ill-conditioning of the stiffness equations. For example, the structural supports can be included in the constraint equations so that the structure as a whole, or components within it, can move as a rigid body. This will make the whole process singular and lead to very unreliable results. Applying the constraints in this manner can be included in a dynamic analysis but it must be incorporated into the eigenvalue extraction algorithm itself which can be inconvenient.

The input data for this form of implementation of constraints generally requires the user to specify each constraint on the form

$$a_0 + a_1 r_1 + a_2 r_2 + a_3 r_3 + \cdots + a_n r_n = 0$$

where r_1 to r_n are the freedoms involved in the constraint and a_0 to a_n are constants that define the constraints. To illustrate the constraint the example shown in Figure 12.9 can be used. In this case the constraint equations are

$$\left(\frac{l_2}{l_1}-1\right)r_1 - \frac{l_2}{l_1}r_2 + r_3 = 0,$$

$$\left(\frac{l_3}{l_1}-1\right)r_1 - \frac{l_3}{l_1}r_2 + r_4 = 0.$$

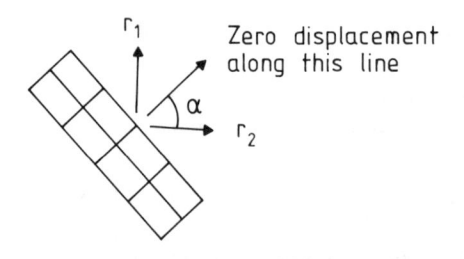

Figure 12.10

This set is not unique and there are various other forms that can be used to express the same constraints. In this example the constraints are not so obvious to define but there are other cases where the equations are more obvious. For example, consider the example of Figure 12.10 where a constraint has to be applied such that the displacement at an angle α is to be zero but the displacement normal to this is free (that is a local coordinate set is to be defined at this node). The constraint can then be written directly as

$$r_1 + r_2 \tan \alpha = 0.$$

The constraints discussed so far are considered to span across elements. A further method that is available in some programs is suitable for those constraints that apply at the element level. The rigid connections discussed above are a typical example of this. Similarly local coordinate directions can be applied at the element level. This form of constraint definition is also very useful for those constraints that cannot be defined directly in terms of the nodal coordinates, typically a constraint that specifies that the material composing the structure is incompressible, or, for fluid flow and acoustic problems, that it is irrotational. These are specified in terms of constraints on the strains and must be included at the element level.

12.8 Comparison of the Forms of Constraint Equations

Few finite element programs contain all possible forms of the constraint equations. Those that are applicable at the element level obviously depend upon the implementation of the elements and are not a general facility. The other two forms of constraints are more general. Programs tend to contain one form or the other. For the most part they are interchangeable and perform the same task. However, they are in fact suitable for different applications and if the user has a choice he should appreciate which technique is most applicable for any given situation. If there are a large number of constraints then the transformation method is better since it reduces the number of equations that have to be solved. The transformation method should also be used where the constraints contain the support definitions so that there are no problems with rigid body movements. The Lagrange multiplier technique is very useful where there are few constraints or where the constraints become active sequentially as the load level increases, typically say in a contact problem. It can also be very useful to correct errors within a mesh that are not discovered until after the loads have been applied. If there are 'cracks' in the mesh because elements are not fully connected, or if compatibility needs to be enforced between some elements, then the fact that the constraints in the Lagrange form are applied at the solution stage means that the user can correct such errors relatively cheaply.

Where there is a choice between the two methods the user can then employ the one that requires the least work on his part in defining the constraints to the system. Some constraints are naturally defined by a transformation equation of the form

$$\mathbf{r} = \mathbf{T}\mathbf{q}$$

where $\mathbf{T}$ is the rectangular matrix defining the constraints. Such a form is natural for example where a cross-section is to be constrained so that it remains plane. In this case the constraint displacements, $\mathbf{q}$, are the transformations and the rotations of the cross-section. $\mathbf{T}$ is then found by considering a unit value for each constraint to be applied in turn and the resulting nodal displacements, $\mathbf{r}$, give the corresponding column in $\mathbf{T}$. The Lagrange form of the constraints, on the other hand, arises naturally when displacements are being equated to each other, typically when defining constraints to enforce compatibility along mesh discontinuities. This point was raised in section 7.3 of chapter 7. Considering the mesh shown in Figure 12.11, there is a discontinuity in mesh density along the boundary

A FINITE ELEMENT PRIMER

marked. If the mesh is analysed as it stands then there will be 'cracks' along this line. Compatibility can be enforced by means of constraint equations. The interpolation function along the common edge for element A for horizontal displacements can be used to give the constraint equations as

$$0.375r_1 - r_2 + 0.75r_3 - 0.125r_5 = 0.$$

$$-0.125r_1 + 0.75r_3 - r_4 + 0.375r_5 = 0.$$

There will also be two very similar equations for compatibility in the vertical direction. Since these equations were derived from the element shape functions they are independent of the actual element geometry and will hence be valid for edges that are not parallel to the global axes. In fact the edge need not even be straight, the only requirement is that the mid-side nodes are actually half way along the arc length.

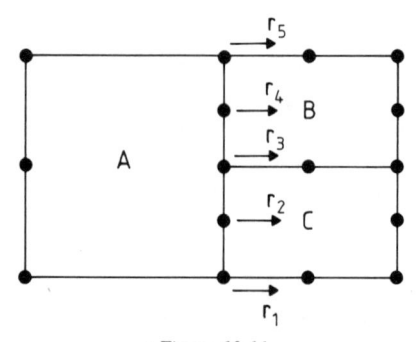

Figure 12.11

12.9 Relationship Between the Forms of Constraint Equations

The two forms of constraint specification are related and constraints given in one form can always be transformed into the other. Typically for the transformation method the constraint equation is

$$\mathbf{r} = \mathbf{T}\mathbf{q}$$

where $\mathbf{T}$ is the $(l \times m)$ rectangular matrix $(l \geq m)$. This can be partitioned into the form

$$\mathbf{T} = \begin{bmatrix} \mathbf{T}_1 \\ \mathbf{T}_2 \end{bmatrix}$$

where T_1 is any partition of T that is a square $(m \times m)$ non-singular matrix. Very often this can be chosen as the unit matrix which considerably simplifies the subsequent operations. T_2 is then an $(l-m) \times m$ rectangular matrix. The Lagrange form of the constraint equations can then be shown to be

$$[-T_2 T_1^{-1} \quad I] \begin{bmatrix} r_1 \\ r_2 \end{bmatrix} = 0$$

where the displacements r have been partitioned to correspond with the partition of T. I is the unit matrix of size $(l-m, l-m)$ so that there are $(l-m)$ constraint equations in this form. Alternatively, the Lagrange form of the constraint equations can be written in the form

$$[A_1 \quad A_2] \begin{bmatrix} r_1 \\ r_2 \end{bmatrix} = 0.$$

If there are n such constraint equations then A_2 is chosen to be a square $(n \times n)$ non-singular partition of this set. The transformation form of the constraint equations can then be written as

$$\begin{bmatrix} r_1 \\ r_2 \end{bmatrix} = \begin{bmatrix} B \\ -A_2^{-1} A_1 B \end{bmatrix} q$$

where B is a square $(n-l, n-l)$ matrix that is used to allow the constraint freedoms, q, to be expressed in some convenient form. In practice it is often taken as the unit matrix in which case $q = r$ and the transformation matrix defines a master/slave relationship.

12.10 Using Mixtures of Element Types

It has been emphasised at various times within this text that the user should take steps to ensure that compatibility is satisfied across elements. If only one type of element is used then, provided that the element shape functions are inherently compatible, there will be complete compatibility throughout the structure. Most systems contain families of elements that are compatible with each other and elements within a compatible family can be mixed together freely. Almost inevitably there are occasions where it is necessary to use elements that are not compatible with each other. As a simple example consider the analysis of the turbine blade shown in Figure 12.12.

A FINITE ELEMENT PRIMER

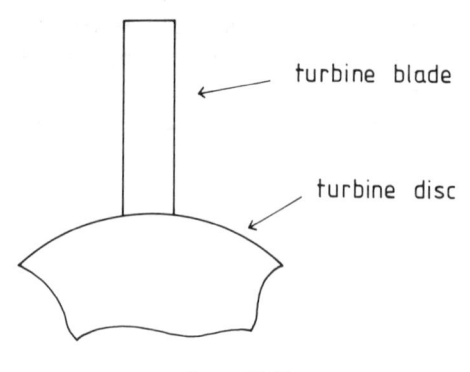

Figure 12.12

The blade itself is most simply modelled as a beam. If the turbine disc is considered to be very stiff then a cantilever beam can be used. In some cases the blade is relatively short compared to its length and shear deflection terms can be included in its description, but the flexibility of the turbine disc can then be important. If this is included by a plate model of the disc there are problems associated with the rotation freedom at the base of the blade. Most plate elements do not have any stiffness associated with in plane rotations, even though they are included as fictitious freedoms. If the user does not do anything about this then the blade will be pinned at its base, at least for in plane rotations. There are various techniques available to enforce compatibility here. Multi-point constraints can be used as illustrated in Figure 12.13.

$$\begin{bmatrix} r_1 \\ r_2 \\ r_3 \\ r_4 \\ r_5 \\ r_6 \\ r_7 \end{bmatrix} = \begin{bmatrix} 1 & 0 & 0 \\ 0 & 1 & -a \\ 1 & 0 & 0 \\ 0 & 1 & a \\ 1 & 0 & 0 \\ 0 & 1 & 0 \\ 0 & 0 & 1 \end{bmatrix} \begin{bmatrix} r_5 \\ r_6 \\ r_7 \end{bmatrix}$$

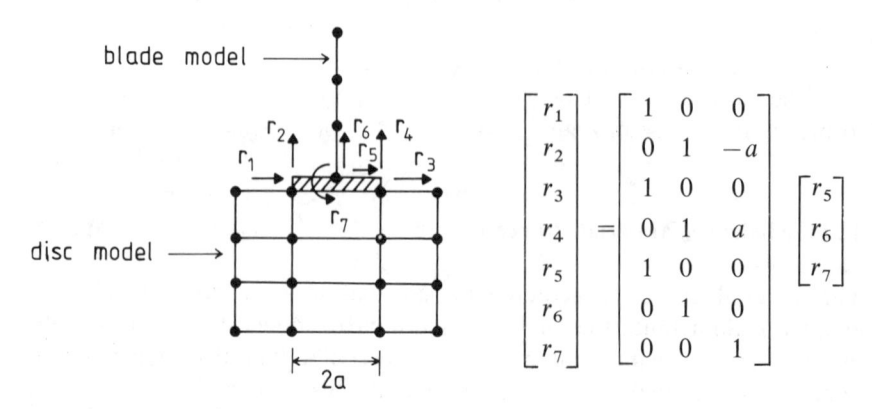

Figure 12.13

This enforces full compatibility for the rotation. Alternatively, the simpler model shown in Figure 12.14 can be used, where the beam elements are extended into the disc for at least three nodes. This will not enforce complete compatibility between the blade and the disc but it will put the end of the blade onto an elastic foundation and will usually give a close approximation to a built in end.

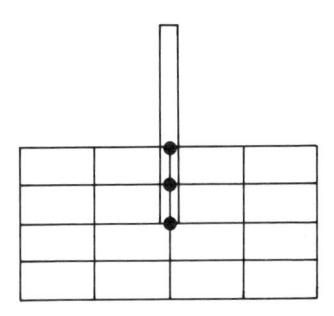

Figure 12.14

Such problems also arise when plate or shell elements are to be connected to three dimensional solids. Again multi-point constraints can be used to achieve full compatibility across the junction. In this case it is better to use constraint forms that involve the transformation method. If the Lagrange constraint equation form is used then a great many constraint equations can be generated and the process can be very expensive. It can also lead to problems in some cases if one or more of the components that are to be connected together are not supported since this will give rise to a singular stiffness matrix. The transformation form of the constraints avoids both of these difficulties. Alternatively, the plate can be extended over the surface (or part of it) to give an approximate connection. These forms are illustrated in Figure 12.15.

12.11 Modelling Material Properties

The material properties required for an analysis have been discussed in chapter 7. Sometimes the analyst is required to model detailed behaviour using a smeared equivalent material property. A typical example of this appears in the analysis of composite materials, where the stress strain relationship (the material stiffness) is used to define the composite form of

A FINITE ELEMENT PRIMER

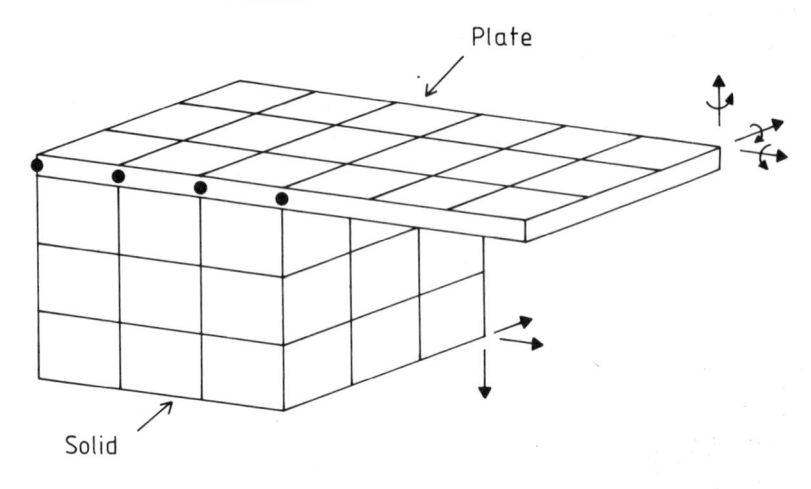

Figure 12.15

the material. In this case there are standard techniques for calculating the equivalent material properties. There are other circumstances in which the user must calculate the equivalent material properties in some manner. For instance, there are many occasions when a structure contains closely spaced and regularly repeated geometrical features. As an example consider the perforated boiler tube plate shown in Figure 12.16.

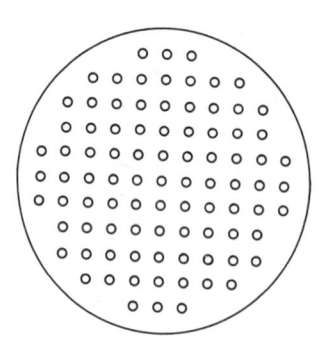

Figure 12.16

There are far too many holes in this to model them in detail but they will have a significant effect upon the structural response of the tube plate. In this case a sufficiently accurate model can be constructed by smearing the effects of the holes throughout the material that composes the tube plate.

252

This will generally lead to a non-isotropic equivalent material model for this portion of the structure. These equivalent material properties can be found by carrying out a detailed analysis of a small portion of the geometry. For the tube plate, away from the edge, conditions of symmetry can be used so that only one quarter of one typical hole need be analysed. A typical two dimensional model is shown in Figure 12.17.

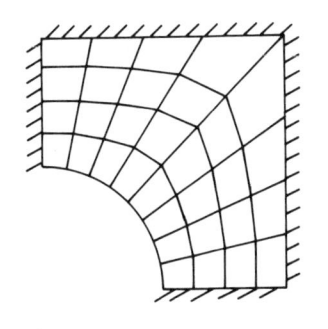

Hatched boundaries are lines of symmetry and are fully fixed

Figure 12.17

Here it is assumed that the through-thickness stresses are zero (the plane stress assumption) but if this is not the case then a full three dimensional model can be used. To determine the equivalent material properties the lines of symmetry are firstly fully fixed. A series of boundary displacements are then applied corresponding to a unit strain in the x-direction, a unit strain in the y-direction and a unit in plane shear as shown in Figure 12.18. For a given applied strain the boundary displacements are r_{bi}. The

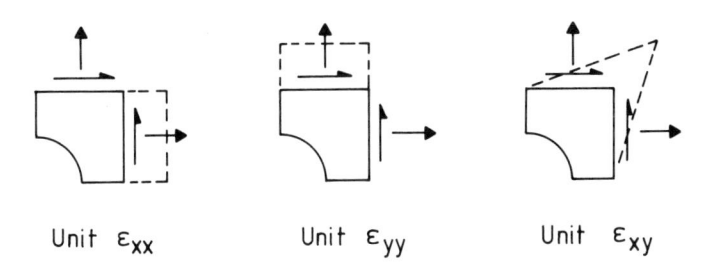

Unit ε_{xx} Unit ε_{yy} Unit ε_{xy}

Arrows show the total reaction forces to be used to determine an equivalent stress - strain law

Figure 12.18

corresponding boundary reaction forces, $\mathbf{R}_{bi}$, are found from the analysis. The equivalent material property matrix, $\mathbf{E}$, can then be found as

$$\mathbf{E} = \frac{1}{\text{vol}} \begin{bmatrix} \mathbf{R}_{b1}^t \\ \mathbf{R}_{b2}^t \\ \mathbf{R}_{b3}^t \end{bmatrix} \begin{bmatrix} \mathbf{r}_{b1} & \mathbf{r}_{b2} & \mathbf{r}_{b3} \end{bmatrix}$$

which gives the correct average strain energy density for the unit strains.

The resulting (3×3) material stiffness matrix can then be used in a plate model for the tube plate, where the holes are not modelled directly and in constructing the mesh the user need not consider the holes at all. He simply divides the plate into the number of elements sufficient to cope with the average stress variations across the structure arising from the overall shape and loadings. The modified material stiffness matrix is fed into the standard (unperforated) elements and the definition of this material matrix is such that the overall effects of the holes are modelled in the full analysis. After this has delivered the strains in the complete model then the detailed stress concentrations due to individual holes can then be found by scaling and summing the stress distributions, found when the material stiffness matrix was formed by multiplying the strains as computed by strain components at the point of interest. This process is not exact since an average value of strain energy was used to compute the equivalent material properties but for a plate with a large number of holes it will give a good representation of the exact behaviour.

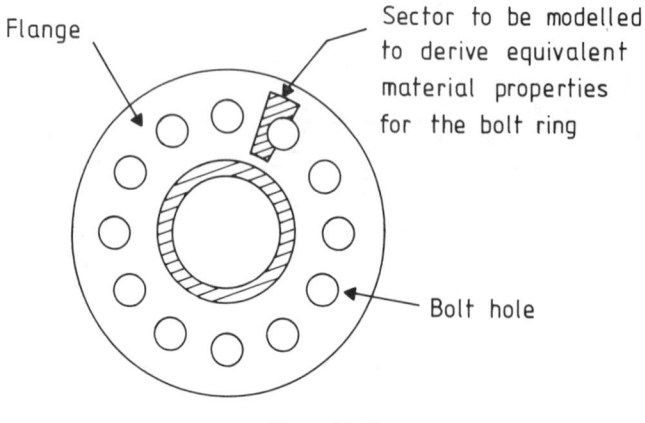

Figure 12.19

A similar process can be used to determine the equivalent stiffness of a row of bolt or rivet holes. Typically an axisymmetric analysis of a pressure vessel usually requires the analyst to model a flange that contains a series of bolt holes around the circumference, as shown in Figure 12.19. These are incompatible with an axisymmetric analysis but a full three dimensional model would be prohibitively expensive. Instead equivalent axisymmetric material properties can be found from the detailed three dimensional analysis of a sector of the flange. This is shown in Figure 12.20, together with the four sets of displacement patterns that are applied to give unit values of strain for each strain component. Again the products of the sets of applied boundary displacements and the corresponding reaction forces can be used to construct the (4×4) equivalent material stiffness matrix for this model. This material stiffness matrix is used for the ring of axisymmetric elements corresponding to the sector model used to derive the equivalent properties. Again detailed stress distributions within any bolt hole can then be recovered from the computed strains components at a hole together with the unit strain cases from the detailed sector model. In this case only one element is needed through the depth of the sector since the behaviour is uniform in the through thickness direction. However, bending or other stress concentration behaviour will be correctly recovered in the axisymmetric model (provided that a sufficiently fine mesh is used).

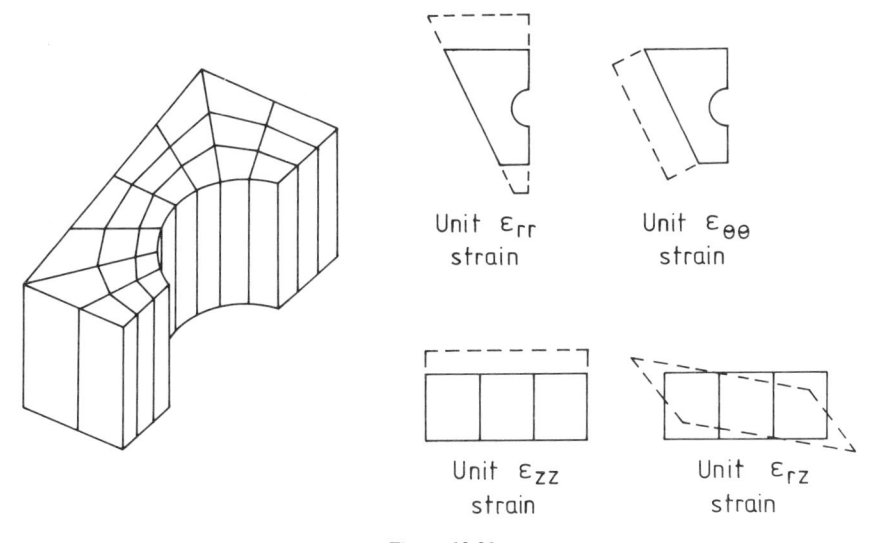

Unit ε_{rr} strain

Unit $\varepsilon_{\theta\theta}$ strain

Unit ε_{zz} strain

Unit ε_{rz} strain

Figure 12.20

A FINITE ELEMENT PRIMER

A similar approach can be used for the analysis of stiffened plates as shown in Figure 12.21. For the equivalent membrane properties a simple assumption is often made that there is zero (or a very small) stiffness in the transverse direction normal to the corrugation but there is longitudinal and shear stiffness. The material Young's modulus in the longitudinal direction is factored by the ratio of the cross-sectional area to the plate cross-sectional area used in the model. A crude smearing idealisation can also be used for the transverse behaviour. The longitudinal flexural stiffness is based upon the actual cross-sectional shape of the plate but the transverse flexural stiffness is based upon the developed length of plate, that is for a flat plate before it is folded into the corrugated shape. It is not easy to see an equivalent simplification for the transverse shear stiffness. This type of simple idealisation is useful where only the displacements are important or where buckling loads are required. There is no consistent way that the analyst can recover the stresses in the plate in this case.

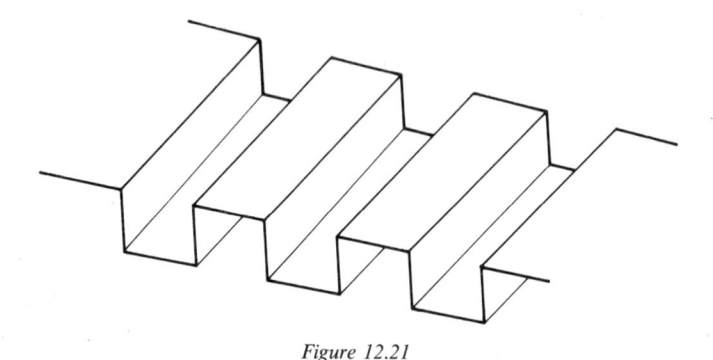

Figure 12.21

In this case a more detailed model of the corrugated plate equivalent material properties can also be derived by the analyst by setting up a detailed model of a small section of the plate. A typical model is shown in Figure 12.22 for a portion of the plate shown in Figure 12.21. Multi-point constraint equations are used to apply unit strains in all of the directions that are relevant for the analysis. The stresses that arise from these unit strains are used to recover the actual stress distribution at any point in the material by multiplying them by the relevant strain components found from the analysis. The use of multi-point constraints is required because the strains must be in a position that corresponds to the mid plane of the equivalent plate used in the model.

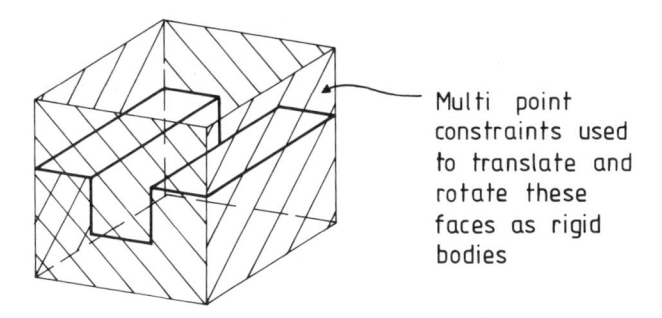

Figure 12.22

This approach to material modelling is also used extensively in the analysis of composite materials. Here the overall bending and stretching behaviour of a thin plate of composite material can be found by considering the sum total of each layer (including offsets) of the composite. This can be done analytically and does not require a finite element model. The individual layer stresses can then be recovered after the overall analysis has been carried out.

Other equivalent material properties can also be found by similar methods. The effective coefficients of thermal expansion can be found by analysing the detailed model under a uniform temperature distribution and calculating the average expansions in each coordinate direction. The method has also proved very useful and efficient in heat conduction problems where equivalent material conductivities, including cross-conductivities, can be found for a fibrous material.

12.12 Modelling of the Loadings

The inherent accuracy of the finite element method is only preserved if care is taken over the definition and modelling of the loadings that are applied in the analysis. Loads can be classified into one of three groups:
a. Applied displacements, where values of the displacements at some points on the structure (usually on the boundaries) are given as initial conditions. A special case of this is support freedoms where the known displacement values are zero.
b. Mechanical loadings, typically point loads, inertia forces, surface pressures and self weight.

c. Initial strain or initial stress loads, typically thermal loads, lack of fit and non-linear material properties.

The applied displacement loadings are found by partitioning the displacement vector into the unknown displacements, r_1, which are being solved for, and the known displacements, r_2, which are given as initial conditions. The forces are partitioned in the same manner into the known forces, R_1, that correspond to the unknown displacements, r_1, and the unknown reaction forces, R_2, that correspond to the known initial displacements, r_2. The stiffness matrix is partitioned to conform to these so that the equations to be solved are

$$\begin{bmatrix} k_{11} & k_{12} \\ k_{21} & k_{22} \end{bmatrix} \begin{bmatrix} r_1 \\ r_2 \end{bmatrix} = \begin{bmatrix} R_1 \\ R_2 \end{bmatrix}.$$

The first equation of this set then gives the equation that is actually solved for the displacements as

$$k_{11} r_1 = R_1 - k_{12} r_2$$

and the reactions can be found by the subsequent operation

$$R_2 = k_{21} r_1 + k_{22} r_2.$$

All of these operations are such that they must be carried out within the computer program itself. The process is not always implemented in a given system in the form described above but any implementation should have the same effect. If it is not available as a standard facility then there is little that the user can do apart from switching to another program.

The mechanical loads are probably the most common form of loading. The finite element method only recognises loads applied at the nodal points and the raw form of loading data consists of specifying a set of nodal loads. A problem then arises for any form of distributed loads, for example a surface pressure, as to how such a load is approximated into equivalent nodal loads. A unique and, as it transpires, the most accurate method of doing this is via kinematically equivalent loads (see equations (4.9) and (5.5)). It is also possible to define a set of statically equivalent loads, where the overall resultant forces and moments are matched between the equivalent and the real loads. For elements with other than the simplest linear interpolation this is not unique and there are any number of statically equivalent load patterns that satisfy equilibrium. To illustrate the superior nature of kinematically equivalent loads consider the

MODELLING

Figure 12.23

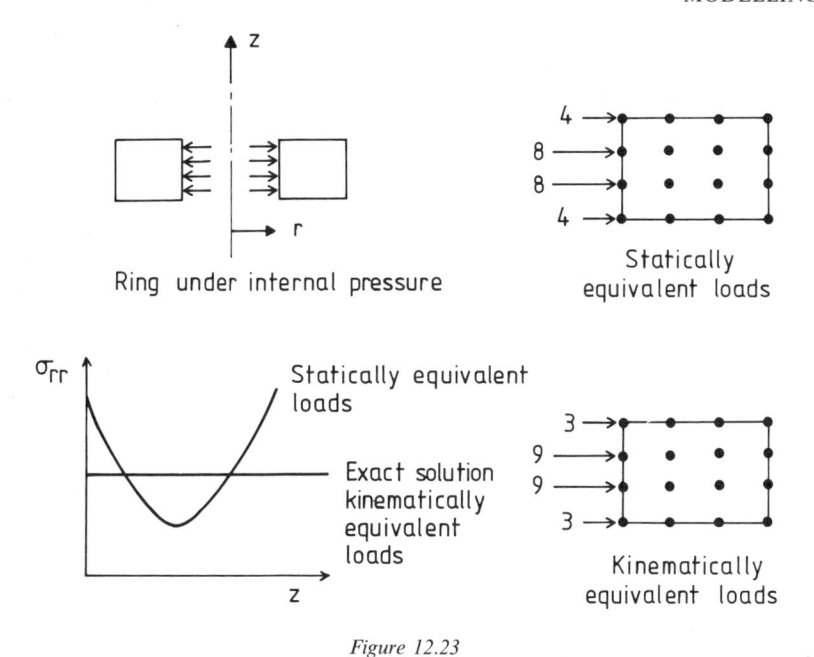

axisymmetric problem shown in Figure 12.23. This is a short cylinder loaded by an internal pressure force p. The exact solution for this demands that all points on the inner surface of the cylinder move radially by the same amount and that the radial stress on the inner surface is equal to the applied pressure. Figure 12.23 shows a comparison with these results for the kinematically equivalent loads and a set of statically equivalent loads that were obtained by lumping the pressure forces on each node on an equal area basis. The analysis was carried out using a third order element (cubic interpolation functions). It will be seen that the kinematically equivalent loads deliver the exact solution at all points on the inner surface. The statically equivalent loads give the correct average values but the distribution is only correct at two points. This is in line with the finite element theory which only guarantees a mean square convergence over the element but it will be appreciated that the kinematically equivalent loads have given a very good point wise convergence in this example. Such a good behaviour is not always achieved but the kinematically equivalent loads will always give better results than any statically equivalent idealisation. Programs should contain facilities for generating kinematically equivalent loads for a range of mechanical loading types. If it does not contain the one that a user

259

A FINITE ELEMENT PRIMER

requires then he can always generate the equivalent loadings external to the FE program and input them as point loads. If the element interpolation functions are known then the kinematically equivalent loads can be formed, otherwise the user must be satisfied with some form of statically equivalent representation. In passing, it is worth noting that any finite element program should print out the overall resultant forces and moments for the mechanical loads that are applied since this gives a very good but simple check that the user has applied the loads correctly.

The third class of loading involves specifying the load as some form of initial strain or initial stress. The more usual form for linear problems is initial strain since this covers thermal and lack-of-fit loads. However, initial stress loadings can be useful to cancel stresses that arise within an element from other loads and thereby put a cut-out in the mesh without actually removing the element or modifying the stiffness matrix in any way. Such loadings differ from mechanical loads in that the initial strains or stresses must be carried through to the stress recovery so that the correct elastic strains and stresses are calculated. The equivalent nodal loads are calculated using the element shape functions. Nodal forces that are found for these loadings must always be self-equilibrating, that is they will have no overall force or moment resultants. It is worth noting that the presence of such loads can make the stress recovery ill-conditioned since it will always involve taking the difference between the total strains and the initial strains in order to find the elastic strains and hence the stresses. In many cases the total and the initial strains are almost equal and care must be taken to ensure that a sufficiently long word length (double precision) is used in order to maintain accuracy. The error that this ill-conditioning induces will not be shown up by the condition number of the stiffness matrix or by a check on overall equilibrium.

12.13 Modelling Considerations Including Loading Effects

There is a great tendency for users to consider that the mesh required for an analysis is defined entirely by the behaviour of the element alone. Whilst this is undoubtedly the most important factor that defines the mesh density it is not the only one. If the loadings are not modelled as accurately as possible then it is very likely that a finer mesh than is really necessary from purely element considerations must be used. This is because most loading idealisations will give rise to oscillatory stress variations throughout an element and large stress variations across elements so that the mean is correct but there is not a pointwise convergence. The user is then forced to specify a fine mesh in order to achieve some degree of point-wise convergence in the stresses. The same effect arises if the program does not allow the user to specify the loadings to the highest degree that the element shape function allows. This is illustrated in Figure 12.24 where the results between two programs are compared for the case of a circular ring loaded by a parabolic temperature variation across the radius of the ring. In tests involving mechanical loads (point loads and radial pressure) the two systems gave very similar results but for the thermal load case one gave much better results than the other. This was traced to the fact that in the accurate one the temperatures at all the eight nodes of each element were included in the load calculation but for the less accurate one only the temperatures at the corner nodes were used. For this example with the mesh as defined this gave a very poor approximation to the actual temperature distribution which was only improved by refining the mesh. Obviously, this gave a higher solution cost. Such considerations are very important for three dimensional analyses where anything that allows the user to use a coarse mesh but still obtain accurate results has a very significant effect on the overall analysis costs.

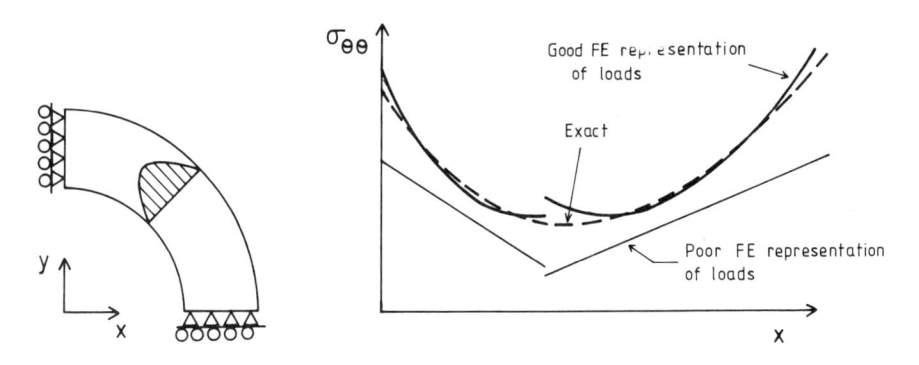

Figure 12.24

13. Other Field Problems

13.1 Introduction

The finite element method is by far the most popular technique for solving structural problems. There is an enormous capital investment in the many commercial finite element systems now in continual use worldwide. It is not surprising therefore that finite element developers should be anxious to use their software in other branches of engineering and science where the same basic programmes can still be used, that is the preprocessor-assembler and analyser-postprocessor. Most current systems will therefore tackle other non-structural problems.

In the early days the finite element method was exclusively structural; and was developed with discrete structures in mind, such as frameworks, multicell tubes, stiffened shells and the like. The ideas and the nomenclature originated in this phase, so the use of 'nodal forces and displacements' stemmed from natural element forces. The main objective of the method was not to satisfy differential equations with boundary conditions, but to assemble a set of physical elements in some way and systematically solve the many simultaneous equations arising from this assembly. The biggest impasse in structural analysis was essentially a geometrical one, and the digital computer helped to solve it.

Contrast this with the world of field equations like fluids and aerodynamics. Traditionally the geometrical shapes involved were simpler than structural frameworks. The designers of transport vehicles, like aircraft or ships, aimed to achieve reasonable flow. Even if separated flow and turbulence were to occur, there was little hope of analysing it. (Not so today.) Therefore it has been true to say that the problems in fluids have tended to concentrate on the nature of the governing equations rather than shape or boundaries. Physical phenomena, like viscosity, compressibility, buoyancy, combustion, cavitation, together with the very large movements of particle flow, led to formidable nonlinear equations

OTHER FIELD PROBLEMS

like the Navier–Stokes (ref. 11). In addition to all this, the equations may possess preferential directions involving convection or wave propagation and these pose special numerical difficulties. Like nonlinear problems in structures there is always the possibility of instability which may be a valid physical instability, like separation, or else quite fictitious numerical instability. Because of the emphasis on solving partial differential equations in field problems, the traditional numerical solution has been to use *finite differences* instead of elements. Meshes were uniform and rarely refined, except perhaps in boundary layers.

So from two extremely different viewpoints, the numerical solution of structures and flow problems has looked to be dissimilar. This is however no longer true; the two approaches have been reconciled, and in some cases may appear together in the same programme. This reconciliation has been brought about by taking a slightly different view of the finite element method, not as a variational problem but as a 'weighted residual problem'. The possible use of structural programmes for any field problem will become immediately clear in this new perspective.

13.2 The Method of Weighted Residuals

We recall the PVD for a general continuum which equated internal to external virtual work,

$$\int_V \sigma^t \bar{\varepsilon} \, dV = \int_V \mathbf{p}_V^t \bar{\mathbf{u}} \, dV + \int_S \mathbf{p}_S^t \bar{\mathbf{u}} \, dS \tag{3.2}$$

and which we showed (after inserting compatibility, $\varepsilon = \partial \mathbf{u}$) led to the form

$$\int_V \bar{\mathbf{u}}^t (\partial^t \sigma + \mathbf{p}_V) \, dV - \int_S \bar{\mathbf{u}}^t [(\partial^t n) \sigma - \mathbf{p}_S] \, dS = 0 \tag{3.15}$$

whence, if the virtual displacements $\bar{\mathbf{u}}$ are continuous and arbitrary everywhere, we concluded that this implied the equilibrium equations in the integrand were equal to zero, $\partial^t \sigma + \mathbf{p}_V = 0$ in V (3.5) and $(\partial^t n) \sigma - \mathbf{p}_S = 0$ in S (3.7).

The finite element trick was to subdivide the integrals in equation (3.2) into summations over elements, and then to make assumptions about the displacement fields which we showed led to (3.5) and (3.7) being satisfied *in the mean* over an element or interface, and not at every point (equations (5.6) and (5.7)).

263

We can therefore view this process entirely in reverse. Suppose we start off with the differential equation and boundary condition to be satisfied. Can we simply take these equations, set to zero, multiply by a suitable (virtual) weight, like $\bar{\mathbf{u}}$, and then write the problem in an integral form like (3.5)? This approach is known as the Method of Weighted Residuals (ref. 12) and has been the object of much study in the past 15 years. It has the overwhelming attraction that we only have to know the differential equation (linear or nonlinear) and boundary condition, and not whether these equations can be extracted from a variational principle. It retains the advantage that we are attempting to satisfy differential equations 'in the mean' only, and that we are free to exploit the nature of the weights. (See ref. 1 for 'upwinding in fluids').

However the MWR is not a panacea, otherwise finite difference codes would now be out of business. It may not always lead to stable algorithms in the case of nonlinear equations, but there is one class of problem for which the MWR not only works well, it also looks like a standard finite element formulation for structures.

Consider the complete set of equations which can be summarised as:

equilibrium	stress-strain	compatibility
$\partial^t \boldsymbol{\sigma} + \mathbf{p}_V = \mathbf{0}$	$\boldsymbol{\sigma} = \mathbf{E}\boldsymbol{\varepsilon}$	$\boldsymbol{\varepsilon} = \partial \mathbf{u}$.

On eliminating stress as a variable we get the following set of equations:

$$\partial^t \mathbf{E} \partial \mathbf{u} + \mathbf{p}_V = \mathbf{0}. \tag{13.1}$$

Differential equations, containing a symmetric operator, of the form $\partial^t \mathbf{E} \partial$, are known as *self adjoint* and we show now that they guarantee a finite element form. One of the most common forms of operator, in all branches of engineering and physics is the *'grad'* operator:

$$\partial = \begin{bmatrix} \dfrac{\partial}{\partial x} \\ \dfrac{\partial}{\partial y} \\ \dfrac{\partial}{\partial z} \end{bmatrix}. \tag{13.2}$$

Inserted into (13.1) this version of ∂ will produce equations known as

OTHER FIELD PROBLEMS

Laplace, Poisson, Harmonic, and more. Instead of treating all of them in general we will look first at the heat conduction problem which contains all the features present in all others. It also happens to be the non-structural finite-element method for which NAFEMS has received the most requests to examine and prepare evaluation programmes.

13.3 The Heat Conduction Problem

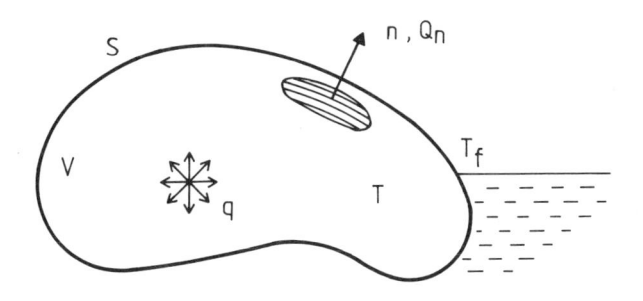

Figure 13.1

Figure 13.1 shows a body with a variable temperature distribution, $T(x, y, z)$, and heat input from several sources. Internally in the volume V there may be heat sources having a flow per unit volume q due to processes for example like chemical, dielectric, nuclear, photosynthetic etc. On the surface there may be a prescribed heat flux Q_n normal to the surface, or heat transfer to a fluid at temperature T_f. Inside V a heat balance across an element leads to

$$\frac{\partial q_x}{\partial x} + \frac{\partial q_y}{\partial y} + \frac{\partial q_z}{\partial z} + q = \rho c \frac{\partial T}{\partial t}$$

where the subscripted 'q' refer to heat flow in those directions, ρ is the density and c the specific heat of the solid. The macroscopic theory of heat conduction due to Fourier states that the heat flux is proportional to temperature gradient, that is

$$q_x = -\kappa_x \frac{\partial T}{\partial x}$$

where κ_x is the thermal conductivity in the 'x' direction. Similar terms

265

A FINITE ELEMENT PRIMER

inserted into the heat flow equation produce

$$\frac{\partial}{\partial x}\left(\kappa_x \frac{\partial T}{\partial x}\right) + \frac{\partial}{\partial y}\left(\kappa_y \frac{\partial T}{\partial y}\right) + \frac{\partial}{\partial z}\left(\kappa_z \frac{\partial T}{\partial z}\right) + q = \rho c \frac{\partial T}{\partial t}.$$

In terms of the gradient operator this can be summarised as:

in V: $\qquad\qquad\qquad \partial^t \kappa \partial T + q = \rho c \dot{T}$ \hfill (13.3)

where κ is the diagonal thermal conductivity matrix

$$\kappa = \begin{bmatrix} \kappa_x & 0 & 0 \\ 0 & \kappa_y & 0 \\ 0 & 0 & \kappa_z \end{bmatrix}.$$

Notice the predicted self-adjoint form of (13.3). On the surface the transfer to the fluid can be expressed in terms of heat transfer coefficient h, according to Newton's law of cooling, so that the total flow is

$$Q_n - h(T - T_f).$$

This is also equal to the emerging flow rate *normal* to the surface, so coupling the three components with their direction cosines $\partial n/\partial x$ etc, we can summarise the boundary conditions as:

on S: $\qquad\qquad (\partial^t n)\kappa \partial T - Q_n + h(T - T_f) = 0.$ \hfill (13.4)

Equations (13.3) and (13.4) contain all the features of the problem, and can be written in weighted residual form, using a *virtual temperature* $\bar{T}(x, y, z)$ as weight thus:

$$\int_V (\partial^t \kappa \partial T + q - \rho c \dot{T}) \bar{T}\, dV = \int_S (\partial^t n)\kappa(\partial T) \bar{T}\, dS. \qquad (13.5)$$

Although a finite element formulation can be created from (13.5) as it stands, it is more convenient to transform to the equivalent of the PVD form by using again a Gauss Identity (3.14):

$$\int_V [(\partial^t T)\kappa(\partial \bar{T}) + (\partial^t \kappa \partial T)\bar{T}]\, dV = \int_S (\partial^t n)\kappa(\partial T)\bar{T}\, dS.$$

OTHER FIELD PROBLEMS

So (13.5) now becomes

$$\int_V (\partial^t T \kappa \partial \bar{T} - q\bar{T} + \rho c \dot{T}\bar{T})\,dV - \int_S [Q_n - h(T-T_f)]\bar{T}\,dS = 0. \quad (13.6)$$

Now we assume that the region is divided into finite elements, each one of which has the temperature interpolated over it by the familiar shape functions

$$T = \mathbf{N}\mathbf{T}_g \quad (13.7)$$

where $\mathbf{T}_g$ is a column of the nodal temperature values. As usual we denote $\partial T = \partial \mathbf{N}\mathbf{T}_g = \mathbf{B}\mathbf{T}_g$, and therefore $\partial^t T = \mathbf{T}_g^t \mathbf{B}^t$ also.

So (13.6) for an element becomes, after taking the transpose,

$$\bar{\mathbf{T}}_g^t [\mathbf{k}_g \mathbf{T}_g + \mathbf{c}_g \dot{\mathbf{T}}_g - \mathbf{P}_g] = 0 \quad (13.8)$$

where the 'element stiffness' $\mathbf{k}_g$ emerges as

$$\mathbf{k}_g = \mathbf{k}_V + \mathbf{k}_S \quad (13.9)$$

and the interior stiffness

$$\mathbf{k}_V = \int_{V_g} \mathbf{B}^t \kappa \mathbf{B}\,dV \quad (13.10)$$

has a stiffness added in (13.9) from the surface (if the element has one) of

$$\mathbf{k}_S = \int_{S_g} \mathbf{N}^t h \mathbf{N}\,dS. \quad (13.11)$$

The element 'damping matrix' comes out as

$$\mathbf{c}_g = \int_{V_g} \mathbf{N}^t \rho c \mathbf{N}\,dV. \quad (13.12)$$

The 'kinematically equivalent loads' appear as

$$\mathbf{P}_g = \int_{V_g} \mathbf{N}^t q\,dV + \int_{S_g} \mathbf{N}^t (Q_n + hT_f)\,dS. \quad (13.13)$$

The element expression is summed as usual, using our symbolic link to the global list of nodal temperatures T, namely $\mathbf{T}_g = \mathbf{a}_g \mathbf{T}$, whence

$$\bar{\mathbf{T}}[\sum \mathbf{a}_g^t \mathbf{k}_g \mathbf{a}_g \mathbf{T} + \sum \mathbf{a}_g^t \mathbf{c}_g \mathbf{a}_g \dot{\mathbf{T}} - \sum \mathbf{a}_g^t \mathbf{P}_g] = 0$$

or for arbitrary $\bar{T}$,

$$\mathbf{KT} + \mathbf{C\dot{T}} = \mathbf{R} \tag{13.14}$$

where $\mathbf{K} = \sum \mathbf{a}_g^t \mathbf{k}_g \mathbf{a}_g$; $\mathbf{C} = \sum \mathbf{a}_g^t \mathbf{c}_g \mathbf{a}_g$; $\mathbf{R} = \sum \mathbf{a}_g^t \mathbf{P}_g$ as usual.

Thus the heat conduction problem simulates a structural problem with stiffness $\mathbf{K}$, damping $\mathbf{C}$, and a load vector $\mathbf{R}$, and the assembly process is no different. However there are subtle differences in detail when we look at the equations.

Firstly T is a scalar, and in (13.5) the virtual weights are $\bar{T}$ in both V and S. So if we imagine (13.5) as a summation over all elements, and provided T is continuous across interfaces, there will be terms from adjacent elements '1' and '2' (see Figure 5.2) like

$$\int_S [(\partial^t n)\kappa \partial T_1 - (\partial^t n)\kappa \partial T_2]\bar{T} dS.$$

Consequently we are attempting to ensure continuity *in the mean* of the heat flux from '1' to '2'. Although this assumes that T is continuous, the gradients ∂T have only to be finite. They can be discontinuous, and mostly are, since only the nodal values of T are our degrees of freedom. The accuracy of this method, and the suitability of the chosen mesh, can be estimated by the discontinuity in normal flux component $(\partial^t n)\kappa(\partial T)$.

Secondly since T is a scalar, no transformation is necessary if we wish to use local axes instead of global, and $\mathbf{k}_g$ does not have to be changed as did structural stiffness (4.13).

Thirdly the element stiffness (13.9) contains a special contribution $\mathbf{k}_S$ if there is a surface, coming from the heat transfer coefficient h. It is usual to treat this term as a separate library 'surface element' of zero thickness, which is assembled via nodal connections in the usual way, and displayed as a surface element during preprocessing checks.

The solution of equation (13.14) in the (static) steady state case $\dot{T}=0$ presents no special difficulties not encountered in structures, provided the right-hand side is constant. However one common case in heat transfer, when this is not so, and $\mathbf{R}$ is a function of T, is when the surface S loses heat by radiation. The heat flux is then proportional to the fourth power of temperature and is written as

$$q = H(T^4 - T_f^4) \quad \text{or} \quad q = h(T - T_f)$$

where

$$h = H(T^2 + T_f^2)(T + T_f). \tag{13.15}$$

This nonlinearity enters the right hand side via hT_f from (13.15) and it also becomes part of the stiffness on the left-hand side of (13.4) via hT: so the problem is now nonlinear. It can be solved by the incremental 'tangent stiffness' methods of chapter 11. It can also be solved by iterating within a single step using the full temperature since the stiffness terms arising from (13.15) are not gradients. Another important nonlinearity may arise if the surrounding fluid does not have a prescribed temperature T_f. There may be buoyant or velocity-dependent convection at the surface of a heat exchanger which makes it necessary to estimate the coupling between the two media.

If we turn to the solution of (13.14) when T_f may be a function of time, or the input Q_n likewise, then the temperature T becomes a variable in time and space. The finite element idealisation takes care of the spatial variation, but all of the nodal values of **T** will vary with time and have to be updated – almost invariably in a numerical incremental fashion.

The techniques for solving dynamic problems in chapter 10 would have us assume a solution of the form

$$\mathbf{T} = \mathbf{T}_0 e^{-wt}$$

so that equation (13.14) – ignoring forcing terms – becomes an eigenvalue problem:

$$(-w\mathbf{C} + \mathbf{K})\mathbf{T}_0 e^{-wt} = 0.$$

Since **C** and **K** are positive definite, the eigenvalues of this set are real and positive, so only 'damped motion' occurs. If a solid is found to respond in any other way, the solution is suspect. On the other hand the prescribed thermal input may be cyclic, so we expect a decaying response leaving the steady state temperature matching the input – but out of phase with it.

When solving the time-varying temperatures it is usual to employ a finite difference scheme on (13.14) in a similar fashion to those discussed in

A FINITE ELEMENT PRIMER

chapter 10 for transient dynamics. The implicit central difference or Crank–Nicholson scheme is used as often as not, in which we form the values of **T** at time $t+\Delta t$ from the values at the beginning of the step t, using

$$(\mathbf{C}+\tfrac{1}{2}\Delta t\mathbf{K})\mathbf{T}_{t+\Delta t}=(\mathbf{C}-\tfrac{1}{2}\Delta t\mathbf{K})\mathbf{T}_t+\tfrac{1}{2}\Delta t(\mathbf{R}_t+\mathbf{R}_{t+\Delta t}). \qquad (13.16)$$

If Δt is kept constant then the inverse $(\mathbf{C}+\tfrac{1}{2}\Delta t\mathbf{K})^{-1}$ has only to be performed once and thereafter $T_{t+\Delta t}$ can be solved in terms of $\mathbf{T}_t$ and $\mathbf{R}_t$ by repeated multiplication. This preferred algorithm is traditionally well behaved, but if the temperature rise time is short (thermal shock) or if the user is tempted to deploy a large increment Δt (which is the same thing) then oscillations may occur and should be damped. Some systems therefore use numerical smoothing techniques on the answers ('laundering') or else introduce a little numerical damping into (13.16) by using coefficients slightly different from the halves.

13.4 Other Field Problems

As we said, the heat conduction problem is the usual facility added to finite element systems. As it happens it contains the most general form of field equation and boundary conditions for this class of self-adjoint problem, and so all others do not require a detailed explanation. We therefore simply list them.

Seepage Flow: $T=$ hydraulic head; $\kappa=$ permeability; $q=0$. This model is viewed with some suspicion by some practitioners of soil mechanics. Soil is a complex medium, and its properties (κ) may depend on the solved pressure anyway (another nonlinear problem).

Potential Flow: $T=$ stream function; $\kappa=[1\ 1\ 1]$; $q=2\times$ vorticity. This is a highly idealised model, ignoring rotational motion and viscosity, and will therefore not cope with boundary layers, separation, and so on. It is useful for estimating pressures, velocity, lift and so on in regions of favourable pressure gradients over smooth surfaces.

Electric Fields: $T=$ voltage; $\kappa=$ electrical conductivity; $q=$ current source. This problem – and the equivalent magnetostatic problem – is usefully solved in the design of motors and the like. There is now much more interest in solving solid-state electronic fields but the simple equations will not then suffice.

OTHER FIELD PROBLEMS

The Torsion Problem and the Membrane: These are largely academic problems, although the torsion problem is of some interest to engineers wishing to find the stiffness, shear centres, and twist of beam sections composed of several component parts. There are specific – not general – finite element programs for this.

References

1. Zienkiewicz, O.C., The Finite Element Method, Third edition, McGraw-Hill, 1977.
2. Davies, G.A.O., Virtual Work in Structural Analysis, John Wiley, 1984.
3. Gallagher, R.H., Finite Element Analysis – Fundamentals, Prentice Hall, 1975.
4. Turner, M.J., Clough, R.W., Martin, H.C. and Topp, L.J., Stiffness and deflection analysis of complex structures, *J. Aero. Sc.* **23**, 805–823, 1956.
5. Henshell, R.D. and Shaw, K.G., Crack tip elements are unnecessary, *Int. Jnl. Num. Meth. Eng.* **9**, 495–509, 1975.
6. Irons, B. and Ahmad, S., Techniques of Finite Elements, Ellis Horwood, 1980.
7. Cook, R.D., Concepts and Applications of Finite Element Analysis, Second edition, John Wiley, 1981.
8. Barlow, J., Optimal stress locations in finite element models, *Int. Jnl. Num. Meth. Eng.* **10**, 243–251, 1976.
9. Irons, B.M., The Semi-Loof shell element, in: Finite Elements for Thin Shells and Curved Members, (eds.) Ashwell & Gallagher, John Wiley, 1976.
10. Hrennikoff, A., Solutions of problems in elasticity by the framework method, *Jnl. App. Mech.* **8**, 1941.
11. Batchelor, G.K., Introduction to Fluid Dynamics, Cambridge University Press, 1967.
12. Finlayson, B.A. and Scriven, L.E., The method of weighted residuals – a review, *App. Mech. Rev.* **19**, 735–748, 1966.

Index

Aircraft construction, 102
angular distortion, 98, 130
area coordinates, 53, 54
aspect ratio distortion, 98, 130
aspect ratios, 76, 129
assembly, 124
assembly of the elements, 3
assembly of stiffness matrix, 15
assembly problems, 127
automatic mesh generation, 86
automatic node renumbering, 149
axisymmetrical element, 71
axisymmetrical shell, 80
axisymmetrical thick shell, 104
axisymmetrical thin shell, 104, 238

Bandwidth, 148
Barlow Points, 63
beam element, 222
beam element stiffness matrix, 36
beams, 29, 32
body force vector, 22
bricks, plates and shells, 68
'bubble' functions, 106, 168
buckling, 176

Central difference method, 270
characteristic values, 174
checking procedures, 126
checks and diagnostics, 145
Cholesky factorisation method, 142, 148
Cholesky solution, 175
code assessments, 157
column vector, 4
compatibility, 9, 30
compatibility equations, 10
compatibility of strains, 14
complete displacement field, 55
composite material, 164, 257
condensation, 196
condition number, 143
conditional stability, 188
conformal elements, 45

congruent transformation, 42
constant strain states, 39
constant stress behaviour, 108
constitutive relationships, 9
constraint equations, 238, 242, 248
contact problems, 101
contour plotting, 167
convergence requirements, 205
convolution integral, 187
corresponding forces and displacements, 5
'crack element', 60
crack propagation, 163
Crank–Nicholson scheme, 270
critical damping, 171, 192
cubic beam nodes, 76
curved 'brick' element, 68
curved Mindlin elements, 82
cyclic symmetry, 116, 122

Damping factor, 171, 193
damping idealisation, 190
damping matrix, 172
decay factor, 171
deformed plotting, 167
degenerate brick elements, 84
diagonal decay, 144
diagonal generalised mass matrix, 174
diagonal generalised stiffness matrix, 174
diagonal mass matrix, 182
die-away length, 95
diffusion, 95
direct integration, 187
direction cosines, 22, 41
discontinuities, 100
discontinuous stiffnesses, 100
discretised approach, 2
displacement method, 12
displacement method, pitfalls of, 51
displacement plots, 156
displacement results, 155
displacement solution, 11
displacement vector, 24
dissimilar material types, 159

INDEX

dissimilar shape functions, 93
distorted quadrilateral elements, 107
distortion, 98, 130
Drucker–Prager equivalent stresses, 164
Duhamel integral, 187
dynamic analysis, 170
dynamic modelling, 180
dynamic stiffness matrix, 200
dynamic stresses, 206
dynamic substructuring, 196, 198

Eigenvalues, 143, 173
eigenvalue problem, 173, 224
eigenvector, 173
elastic foundation, 242
elastic stiffness, 220
electric fields, 270
element assembly, 3
element distortion, 97, 128
element generation, 127
element Jacobian, 98
element stiffness, 17, 18, 19, 44
element strains, 164
element stresses, 158
element type, 102
elliptical hole, 60
energy method, 13
engineering normalisation, 174
equations of compatibility, 24
equilibrium, 9
equilibrium equations, 10
equilibrium finite elements, 51
equivalent stres, 162
equivalent material properties, 252
Eulerian method, 221
exact solutions, 29
explicit methods, 189
external potential energy, 15
extrapolation, 161

Faceted geometry, 89
fail safe case, 157
fault conditions, 157
fictitious very stiff elements, 101
field problems, 262
finite differences, 263
flat thin shell elements, 105
flexibility, 6
flexibility matrix, 6
flow chart, 124
Force method, 155
forced response, 184, 186
forcing functions, 172
forms of constraint equation, 247
Fourier expansions, 83
Fourier series, 72, 121

Fourier transform, 186, 205, 212
Framework Analysis, 15
free vibrations, 173
frequency domain, 186
frontal solution, 149
fundamental work equation, 27

Gap elements, 101
Gauss Point stresses, 64, 230
Gauss Point extrapolation, 158
Gauss Points, 62
Gauss' Theorem, 27, 45
Gaussian elimination, 141
Gaussian integration, 61
Gaussian Quadrature, 61
general continuum, 22
general thin shell, 104
geometric stiffness, 220, 223
geometrical errors, 88
geometry specification, 86
global stiffness matrix, 17, 37
global stresses, 158
'grad' operator, 264
graphical mesh checks, 133
graphical presentation, 166
graphics requirements, 133
gross deformations, 217
guard vectors, 178
Guyan reduction method, 197
gyroscopic forces, 194

Hardening structure, 226
harmonic loading, 73
heat conduction, 265
Hermitian shape functions, 83
hidden line removal, 134
high aspect ratio, 129
Hooke's Law, 11
'hourglass' mode, 70
hybrid elements, 52, 155
hysteretic damping, 195

Ill-conditioned equations, 101
ill-conditioning, 40, 140, 179, 245, 260
ill-conditioning errors, 165
implicit methods, 189
incremental solutions, 224
inelastic material behaviour, 227
initial buckling, 223
incorrect element connections, 131
incorrect units, 112
initial strains, 65
integration by parts, 27
internal virtual work, 26
interpolation functions, 33
isoparametric curved quad, 56

INDEX

isoparametric element, 56
isotropic material, 25

Jacobi method, 176
Jacobian, 98, 100, 129, 160
Jacobian matrix, 58
J-integral methods, 163
joints, 234

Kinematic boundary conditions, 31
kinematic connectivity, 37
kinematic mechanism, 70
kinematically equivalent forces, 45, 66
kinematically equivalent loads, 258
kinematically equivalent nodal forces, 36, 165

Lagrange multipliers, 245
Lagrange multiplier technique, 247
Lagrangian method, 221
Lanczos method, 178
latent roots, 174
least squares fit, 66
linear dependence, 82
lines of intersection, 89
lines of principle curvature, 84
loadings, 257
local stresses, 158
lower bound, 34
lower bound solution, 34
lumped mass form, 182

Mass matrix, 172
'master' freedoms, 197
material flexibility matrix, 113
material loss factor, 195
material properties, 112, 251
material stiffness, 25
material stiffness matrix, 113
mathematical normalisation, 174
matrix array, 6
matrix inverse, 8
matrix notation, 4
matrix products, 7
Matrix Force Method, 15
matrix transpose, 5
mean square convergence, 161
mean square values, 213
membrane behaviour, 75
membranes, 43
mesh density, 94, 106
mesh generation, 86, 90
mesh refinement, 168
mesh specification, 85
mesh suitability, 105
mid-node position distortion, 98

Mindlin model, 77
mirror image symmetries, 116
mixing of element types, 132, 249
modal condensation, 197
modal damping, 191
mode participation factor, 204
modelling, 232
modified material stiffness, 254
modified Newton–Raphson, 226
Mohr–Coulomb equivalent stress, 164
multi-point constraints, 93, 250

NAFEMS benchmarks, 60, 66
natural frequency, 170
Navier–Stokes equations, 263
Neumark method, 189
Newton–Cotes formulae, 62
Newton–Raphson, 226
Newton–Raphson, modified, 226
nodal values, 33
nodes, 33
'non-conforming' elements, 46
non-linear analysis, 153, 215
non-linear problems, 223
non-stationary random vibrations, 210, 185
non-structural mass, 182
norm, 145
normal modes, 174
normalisation, 174
nozzle/cylinder intersection, 139
number of modes (for a dynamic analysis), 202
numerical integration, 61

Offsets, 235
Offset problems, 243
optimal sampling points, 63
outline plotting, 134
overstiff solutions, 34

Parametric studies, 110
'Patch Test', 46, 102, 107
percentage critical damping, 192
physical discontinuities, 100
pipework system, 138
plane strain assumption, 114
plane stress material stiffness, 114
plastic flexibility, 228
plate bending elements, 75
Poisson's ratio, 25
post analysis checks, 127
post-processing, 157, 147, 157
potential flow, 270
Power methods, 175
Prandtl–Reuss flow rule, 229
Prandtl–Reuss theory, 228

INDEX

primary component, 200
principal curvature, 84
principal stresses, 162
profile of equation minimisation, 148
program restarts, 151
proportional damping, 191
Principle of Virtual Displacements, 14, 27
PVD for solid, 26

QR method, 176
quality assurance, 154

Radiation, 268
radiation damping, 191
random vibrations, 212
rank 'deficiency', 20
rank of a matrix, 20
Rayleigh damping, 191
Rayleigh quotient, 180, 206
reaction forces, 165
rectangular elements, 47
reduced integration, 48, 63, 69, 148, 155
renumbering, 150
repetitive symmetry, 116, 123
response spectrum, 205
response spectrum method, 211
restarts, 124
results presentation, 146
results processing, 154
rigid body modes, 38
rigid body movements, 107
rigid jointed frameworks, 39
roundoff error, 141, 143
row x column rule, 5, 6
row matrix, 5

Scalars, 7
secant stiffness, 225
secondary components, 200
seepage flow, 270
seismic analysis, 210
selected reduced integration, 83
self adjoint equations, 264
Semi-Loof element, 83
'shakedown', 230
shape functions, 33, 45, 47, 48, 54
shell element, thick, 104
shell element, thin, 238
shell elements, 79
Simpson's Rule, 32, 62
simultaneous vector iteration, 177
singular matrix, 19
singular stiffness matrix, 69, 240
skew distortion, 129
skyline storage, 149
'slave' freedoms, 197

snap through, 218
solid elements, 68
solution diagnostics, 141
solution efficiency, 147
sparse matrix methods, 177
spectral density, 212
spline curves, 88, 89
spurious cracks, 87
statically determinate structure, 12
statically equivalent loads, 258
statically indeterminate structure, 12
stationary random excitation, 185
step-by-step integration, 187, 209
stiff elements, 101
stiffness matrix, 8
stiffness matrix, global, 17, 37
stiffness transformation, 40
strain energy, 15
strain vector, 23
stress averaging, 159, 160
stress concentration, 95, 206
stress contours, 162
stress deviators, 229
stress discontinuities, 159
stress extrapolation, 158
stress intensity factors, 152, 163
stress presentation, 162
stress results, 156
stress smoothing, 161
stress vector, 23
stress–strain law, 9
structural modifications, 208
Sturm sequence check, 179
subspace vector iteration, 177
substructuring, 152
super element method, 153
supports, 239
surface element, 268
symmetrical matrix, 8
symmetry, use of, 115

Tangent stiffness, 220, 224, 269
taper distortion, 129
tetrahedron, 70
thermal conductivity matrix, 266
thermal loads, 155, 261
thermal strains, 25
time domain, 186
torsion problem, 271
trace of the matrix, 144
transformation method, 176
transient force, 184
transition element, 92
Tresca equivalent stress, 163
triangular elements, 52
top down approach, 111
turbine blades and disc, 250

Ultimate stress, 157
unconditional stability, 188
unknown redundancies, 12
unit matrix, 41
updated Lagrangian, 222
'upwinding in fluids', 264

Variable bandwidth, 149
variable mesh density, 96
velocity response, 205
violating compatibility, 90
virtual crack extensions, 152, 163
virtual displacements, 14
Virtual Displacements, Principle of, 13, 14, 27
Virtual Forces, Principle of, 15
virtual work, 13, 26, 44, 222
Virtual Work, Principle of, 13

viscous damping matrix, 190
volume distortion, 129
volumetric distortion, 98
Von Mises equivalent stress, 163
Von Mises reference stress, 229

Wave propagation, 209
weighted residuals, 263
whirling stability, 195
Wilson-θ method, 188
window command, 167
word length, 140

Young's Modulus, 25

Zero energy mode, 70
zero stiffness modes, 148